Business Process Management

Business Process Management

Edited by
Fabian McLain

Larsen & Keller
www.larsen-keller.com

Business Process Management
Edited by Fabian McLain
ISBN: 978-1-63549-234-7 (Hardback)

Larsen & Keller
Published by Larsen and Keller Education,
5 Penn Plaza,
19th Floor,
New York, NY 10001, USA

Cataloging-in-Publication Data

Business process management / edited by Fabian McLain.
p. cm.
Includes bibliographical references and index.
ISBN 978-1-63549-234-7
1. Workflow--Management. 2. Business planning. 3. Industrial management.
I. McLain, Fabian.
HD62.17 .B87 2017
658.401--dc23

Printed and bound in China.

For more information regarding Larsen and Keller Education and its products, please visit the publisher's website www.larsen-keller.com

Table of Contents

	Preface	**VII**
Chapter 1	**Introduction to Business Process Management**	**1**
	i. Business Process Management	1
	ii. Operations Management	9
	iii. Management	28
Chapter 2	**Business Model and its Types**	**43**
	i. Business Model	43
	ii. Bricks and Clicks	54
	iii. Business Model Pattern	58
	iv. Loyalty Business Model	58
	v. Industrialization of Services Business Model	63
	vi. Dependent Growth Business Model	64
	vii. Pyramid Scheme	64
	viii. Business Models for Open-source Software	70
	ix. Inclusive Business Model	80
Chapter 3	**Business Process Modeling: An Overview**	**87**
	i. Business Process Modeling	87
	ii. Business Reference Model	93
	iii. Business Process Model and Notation	96
	iv. Business Process	109
	v. Artifact-centric Business Process Model	117
	vi. Event-driven Process Chain	119
	vii. Disintermediation	125
	viii. Direct Selling	128
	ix. Freemium	130
Chapter 4	**Diverse Aspects of Business Process Management**	**135**
	i. Business Process Reengineering	135
	ii. Outsourcing	149
	iii. Business Activity Monitoring	162
	iv. Collective Business System	164
	v. Internet of Things	166
	vi. Business Process Automation	189
	vii. Business Process Customization	191
	viii. Business Process Preservation	192
	ix. Professional Open Source	193
	x. Virtual Business Model	197

Chapter 5 Business Intelligence: An Integrated Study **199**
i. Business Intelligence 199
ii. Business Intelligence 2.0 211
iii. Business Analytics 211
iv. Analytics 214
v. Metrics Reference Model 218
vi. Financial Intelligence (Business) 222
vii. IBM Cognos Business Intelligence 224
viii. Report 227
ix. Collaborative Software 229
x. Text Mining 235

Permissions

Index

Preface

Business process management refers to that branch of operations management which deals with the management and optimization of a firm's business processes. It is used to make organizations more capable, efficient and effective to change. This book provides comprehensive insights into the field of business process management. It attempts to understand the multiple branches that fall under this discipline and how such concepts have practical applications. Such selected concepts that redefine this subject area have been presented in it. Different approaches, evaluations and methodologies have been included in it. This textbook is a complete source of knowledge on the present status of this important field.

A short introduction to every chapter is written below to provide an overview of the content of the book:

Chapter 1 - Business process management focuses on improving corporate performances. It can also be described as a process optimization process. Business process management helps organizations to be more efficient and also more effective. The chapter on business process management offers an insightful focus, keeping in mind the subject matter; **Chapter 2** - Business models are models that are used as abstracts to represent an organization. It is used to describe the procedure of how an organization is created and delivered. The process of contrasting this model is a part of business strategy. The topics discussed in the chapter are of great importance to broaden the existing knowledge on business models; **Chapter 3** - Business process modeling is the activity used in systems engineering. The use of this activity is to represent processes of any enterprise. Some aspects of business process modeling elucidated in this section are business reference model, business process and direct selling. This chapter is an overview of the subject matter incorporating all the major aspects of business process modeling; **Chapter 4** - The aspects of business process management explained in the following section are business process reengineering, outsourcing, business activity monitoring, Internet of things, business process automation and virtual business model. Business process reengineering is a management strategy that focuses and analysis workflows and the process within any organization. The major components of business process management are discussed in this chapter; **Chapter 5** - Business intelligence is a theory that includes the strategies and applications and also the technologies that are used in any business. It helps in easy interpretations of large volumes of data. The aspects elucidated in this section are of vital importance, and provides a better understanding of business intelligence.

I extend my sincere thanks to the publisher for considering me worthy of this task. Finally, I thank my family for being a source of support and help.

Editor

1

Introduction to Business Process Management

Business process management focuses on improving corporate performances. It can also be described as a process optimization process. Business process management helps organizations to be more efficient and also more effective. The chapter on business process management offers an insightful focus, keeping in mind the subject matter.

Business Process Management

Business process management (BPM) is a field in operations management that focuses on improving corporate performance by managing and optimizing a company's business processes. It can therefore be described as a "process optimization process." It is argued that BPM enables organizations to be more efficient, more effective and more capable of change than a functionally focused, traditional hierarchical management approach. These processes can impact the cost and revenue generation of an organization.

As a policy-making approach, BPM sees processes as important assets of an organization that must be understood, managed, and developed to announce value-added products and services to clients or customers. This approach closely resembles other total quality management or continual improvement process methodologies and BPM proponents also claim that this approach can be supported, or enabled, through technology. As such, many BPM articles and scholars frequently discuss BPM from one of two viewpoints: people and/or technology.

Definitions

BPMInstitute.org defines Business Process Management as:

> the definition, improvement and management of a firm's end-to-end enterprise business processes in order to achieve three outcomes crucial to a performance-based, customer-driven firm: 1) clarity on strategic direction, 2) alignment of the firm's resources, and 3) increased discipline in daily operations.

The Workflow Management Coalition, BPM.com and several other sources have come to agreement on the following definition:

> Business Process Management (BPM) is a discipline involving any combination of modeling, automation, execution, control, measurement and optimization of business activity flows, in support of enterprise goals, spanning systems, employees, customers and partners within and beyond the enterprise boundaries.

The Association Of Business Process Management Professionals defines BPM as:

> Business Process Management (BPM) is a disciplined approach to identify, design, execute, document, measure, monitor, and control both automated and non-automated business processes to achieve consistent, targeted results aligned with an organization's strategic goals. BPM involves the deliberate, collaborative and increasingly technology-aided definition, improvement, innovation, and management of end-to-end business processes that drive business results, create value, and enable an organization to meet its business objectives with more agility. BPM enables an enterprise to align its business processes to its business strategy, leading to effective overall company performance through improvements of specific work activities either within a specific department, across the enterprise, or between organizations.

Gartner defines Business process management (BPM) as:

> "the discipline of managing processes (rather than tasks) as the means for improving business performance outcomes and operational agility. Processes span organizational boundaries, linking together people, information flows, systems and other assets to create and deliver value to customers and constituents."

It is common to confuse BPM with a BPM Suite (BPMS). BPM is a professional discipline done by people, whereas a BPMS is a technological suite of tools designed to help the BPM professionals accomplish their goals. BPM should also not be confused with an application or solution developed to support a particular process. Suites and solutions represent ways of automating business processes, but automation is only one aspect of BPM.

Changes in Business Process Management

The concept of business process may be as traditional as concepts of tasks, department, production, and outputs, arising from job shop scheduling problems in the early 20th Century. The management and improvement approach as of 2010, with formal definitions and technical modeling, has been around since the early 1990s. Note that the term "business process" is sometimes used by IT practitioners as synonymous with the management of middleware processes or with integrating application software tasks.

Although BPM initially focused on the automation of business processes with the use of information technology, it has since been extended to integrate human-driven processes in which human interaction takes place in series or parallel with the use of technology. For example, workflow management systems can assign individual steps requiring deploying human intuition or judgment to relevant humans and other tasks in a workflow to a relevant automated system.

More recent variations such as "human interaction management" are concerned with the interaction between human workers performing a task.

As of 2010 technology has allowed the coupling of BPM with other methodologies, such as Six Sigma. Some BPM tools such as SIPOCs, process flows, RACIs, CTQs and histograms allow users to:

- visualize - functions and processes
- measure - determine the appropriate measure to determine success
- analyze - compare the various simulations to determine an optimal improvement
- improve - select and implement the improvement
- control - deploy this implementation and by use of user-defined dashboards monitor the improvement in real time and feed the performance information back into the simulation model in preparation for the next improvement iteration
- re-engineer - revamp the processes from scratch for better results

This brings with it the benefit of being able to simulate changes to business processes based on real-world data (not just on assumed knowledge). Also, the coupling of BPM to industry methodologies allows users to continually streamline and optimize the process to ensure that it is tuned to its market need.

As of 2012 research on BPM has paid increasing attention to the compliance of business processes. Although a key aspect of business processes is flexibility, as business processes continuously need to adapt to changes in the environment, compliance with business strategy, policies and government regulations should also be ensured. The compliance aspect in BPM is highly important for governmental organizations. As of 2010 BPM approaches in a governmental context largely focus on operational processes and knowledge representation. Although there have been many technical studies on operational business processes in both the public and private sectors, researchers have rarely taken legal compliance activities into account, for instance the legal implementation processes in public-administration bodies.

BPM Life-cycle

Business process management activities can be arbitrarily grouped into categories such as design, modeling, execution, monitoring, and optimization.

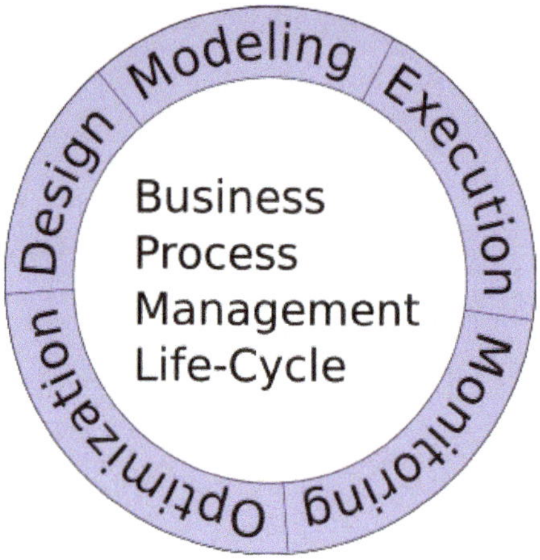

Design

Process design encompasses both the identification of existing processes and the design of "to-be" processes. Areas of focus include representation of the process flow, the factors within it, alerts and notifications, escalations, standard operating procedures, service level agreements, and task hand-over mechanisms.

Whether or not existing processes are considered, the aim of this step is to ensure that a correct and efficient theoretical design is prepared.

The proposed improvement could be in human-to-human, human-to-system or system-to-system workflows, and might target regulatory, market, or competitive challenges faced by the businesses.

The existing process and the design of new process for various applications will have to synchronise and not cause major outage or process interruption.

Modeling

Modeling takes the theoretical design and introduces combinations of variables (e.g., changes in rent or materials costs, which determine how the process might operate under different circumstances).

It may also involve running "what-if analysis"(Conditions-when, if, else) on the processes: *"What if I have 75% of resources to do the same task?" "What if I want to do the same job for 80% of the current cost?"*.

Execution

One of the ways to automate processes is to develop or purchase an application that executes the required steps of the process; however, in practice, these applications rarely execute all the steps of the process accurately or completely. Another approach is to use a combination of software and human intervention; however this approach is more complex, making the documentation process difficult.

As a response to these problems, software has been developed that enables the full business process (as developed in the process design activity) to be defined in a computer language which can be directly executed by the computer. The process models can be run through execution engines that automate the processes directly from the model *(e.g.* calculating a repayment plan for a loan) or, when a step is too complex to automate, Business Process Modeling Notation (BPMN) provides front-end capability for human input. Compared to either of the previous approaches, directly executing a process definition can be more straightforward and therefore easier to improve. However, automating a process definition requires flexible and comprehensive infrastructure, which typically rules out implementing these systems in a legacy IT environment.

Business rules have been used by systems to provide definitions for governing behavior, and a business rule engine can be used to drive process execution and resolution.

Monitoring

Monitoring encompasses the tracking of individual processes, so that information on their state can be easily seen, and statistics on the performance of one or more processes can be provided. An example of this tracking is being able to determine the state of a customer order *(e.g.* order arrived, awaiting delivery, invoice paid) so that problems in its operation can be identified and corrected.

In addition, this information can be used to work with customers and suppliers to improve their connected processes. Examples are the generation of measures on how quickly a customer order is processed or how many orders were processed in the last month. These measures tend to fit into three categories: cycle time, defect rate and productivity.

The degree of monitoring depends on what information the business wants to evaluate and analyze and how business wants it to be monitored, in real-time, near real-time or ad hoc. Here, business activity monitoring (BAM) extends and expands the monitoring tools generally provided by BPMS.

Process mining is a collection of methods and tools related to process monitoring. The aim of process mining is to analyze event logs extracted through process monitoring and to compare them with an *a priori* process model. Process mining allows process analysts to detect discrepancies between the actual process execution and the *a priori* model as well as to analyze bottlenecks.

Optimization

Process optimization includes retrieving process performance information from modeling or monitoring phase; identifying the potential or actual bottlenecks and the potential opportunities for cost savings or other improvements; and then, applying those enhancements in the design of the process. Process mining tools are able to discover critical activities and bottlenecks, creating greater business value.

Re-engineering

When the process becomes too complex or inefficient, and optimization is not fetching the desired output, it is usually recommended by a company steering committee chaired by the president / CEO to re-engineer the entire process cycle. Business process reengineering (BPR) has been used by organizations to attempt to achieve efficiency and productivity at work.

BPM Suites

A market has developed for Enterprise software leveraging the Business Process Management concepts to organize and automate processes. The recent convergence of these software from distinct pieces such as Business rules engine, Business Process Modelling, Business Activity Monitoring and Human Workflow has given birth to integrated Business Process Management Suites. Forrester Research, Inc recognize the BPM suite space through three different lenses:

- human-centric BPM
- integration-centric BPM (Enterprise Service Bus)
- document-centric BPM (Dynamic Case Management)

However, standalone integration-centric and document-centric offerings have matured into separate, standalone markets.

Practice

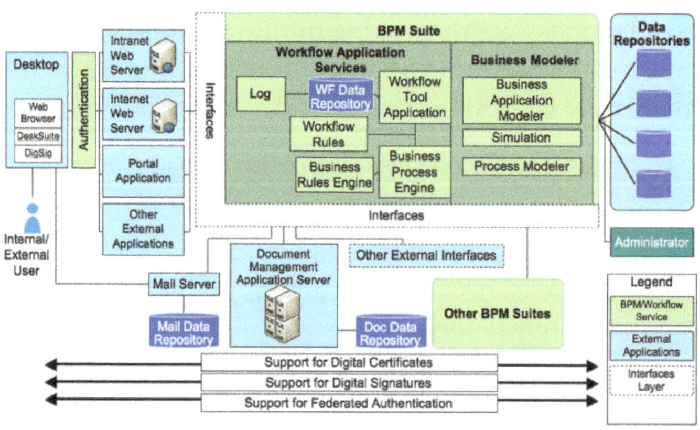

Example of Business Process Management (BPM) Service Pattern: This pattern shows how business process management (BPM) tools can be used to implement business processes through the orchestration of activities between people and systems.

While the steps can be viewed as a cycle, economic or time constraints are likely to limit the process to only a few iterations. This is often the case when an organization uses the approach for short to medium term objectives rather than trying to transform the organizational culture. True iterations are only possible through the collaborative efforts of process participants. In a majority of organizations, complexity will require enabling technology to support the process participants in these daily pro-cess management challenges.

To date, many organizations often start a BPM project or program with the objective of optimizing an area that has been identified as an area for improvement.

Currently, the international standards for the task have limited BPM to the application in the IT sector, and ISO/IEC 15944 covers the operational aspects of the business. However, some corporations with the culture of best practices do use standard operating procedures to regulate their operational process. Other standards are currently being worked upon to assist in BPM implementation (BPMN, Enterprise Architecture, Business Motivation Model).

BPM Technology

BPM is now considered a critical component of operational intelligence (OI) solutions to deliver real-time, actionable information. This real-time information can be acted upon in a variety of ways - alerts can be sent or executive decisions can be made using real-time dashboards. OI solutions use real-time information to take automated action based on pre-defined rules so that security measures and or exception management processes can be initiated.

As such, some people view BPM as "the bridge between Information Technology (IT) and Business.". In fact, an argument can be made that this "holistic approach" bridges organizational and technological silos.

There are four critical components of a BPM Suite:

- Process engine — a robust platform for modeling and executing process-based applications, including business rules
- Business analytics — enable managers to identify business issues, trends, and opportunities with reports and dashboards and react accordingly
- Content management — provides a system for storing and securing electronic documents, images, and other files
- Collaboration tools — remove intra- and interdepartmental communication barriers through discussion forums, dynamic workspaces, and message boards

BPM also addresses many of the critical IT issues underpinning these business drivers, including:

- Managing end-to-end, customer-facing processes
- Consolidating data and increasing visibility into and access to associated data and information
- Increasing the flexibility and functionality of current infrastructure and data
- Integrating with existing systems and leveraging service oriented architecture (SOA)
- Establishing a common language for business-IT alignment

Validation of BPMS is another technical issue that vendors and users need to be aware of, if regulatory compliance is mandatory. The validation task could be performed either by an authenticated third party or by the users themselves. Either way, validation documentation will need to be generated. The validation document usually can either be published officially or retained by users.

Cloud Computing BPM

Cloud computing business process management is the use of (BPM) tools that are delivered as software services (SaaS) over a network. Cloud BPM business logic is deployed on an application server and the business data resides in cloud storage.

Market

According to Gartner, 20% of all the "shadow business processes" will be supported by BPM cloud platforms. Gartner refers to all the hidden organizational processes that are supported by IT departments as part of legacy business processes such as Excel spreadsheets, routing of emails using rules, phone calls routing, etc. These can, of course also be replaced by other technologies such as workflow software.

Benefits

The benefits of using cloud BPM services include removing the need and cost of maintaining specialized technical skill sets in-house and reducing distractions from an enterprise's main focus. It offers controlled IT budgeting and enables geographical mobility..

Internet of Things

The emerging Internet of Things poses a significant challenge to control and manage the flow of information through large numbers of devices. To cope with this, a new direction known as BPM Everywhere shows promise as way of blending traditional process techniques, with additional capabilities to automate the handling of all the independent devices.

Operations Management

Operations management is an area of management concerned with designing and controlling the process of production and redesigning business operations in the production of goods or services. It involves the responsibility of ensuring that business operations are efficient in terms of using as few resources as needed and effective in terms of meeting customer requirements. It is concerned with managing the process that converts inputs (in the forms of raw materials, labor, and energy) into outputs (in the form of goods and/or services). The relationship of operations management to senior management in commercial contexts can be compared to the relationship of line officers to highest-level senior officers in military science. The highest-level officers shape the strategy and revise it over time, while the line officers make tactical decisions in support of carrying out the strategy. In business as in military affairs, the boundaries between levels are not always distinct; tactical information dynamically informs strategy, and individual people often move between roles over time.

Ford Motor car assembly line: the classical example of a manufacturing production system.

Post office queue. Operations management studies both manufacturing and services.

According to the United States Department of Education, operations management is the field concerned with managing and directing the physical and/or technical functions of a firm or organization, particularly those relating to development, production, and manufacturing. Operations management programs typically include instruction in principles of general management, manufacturing and production systems, factory management, equipment maintenance management, production control, industrial labor relations and skilled trades supervision, strategic manufacturing policy, systems analysis, productivity analysis and cost control, and materials planning.

History

The history of production and operation systems began around 5000 B.C. when Sumerian priests developed the ancient system of recording inventories, loans, taxes, and business transactions. The next major historical application of operation systems occurred in 4000 B.C. It was during this time that the Egyptians started using planning, organization, and control in large projects such as the construction of the pyramids. By 1100 B.C., labor was being specialized in China; by about 370 B.C., Xenophon described the advantages of dividing the various operations necessary for the production of shoes among different individuals in ancient Greece.

Shoemakers, 1568

In the Middle Ages, kings and queens ruled over large areas of land. Loyal noblemen maintained large sections of the monarch's territory. This hierarchical organization in which people were divided into classes based on social position and wealth became known as the feudal system. In the feudal system, vassals and serfs produced for themselves and people of higher classes by using the ruler's land and resources. Although a large part of labor was employed in agriculture, artisans contributed to economic output and formed guilds. The guild system, operating mainly between 1100 and 1500, consisted of two types: merchant guilds, who bought and sold goods, and craft guilds, which made goods. Although guilds were regulated as to the quality of work performed, the resulting system was rather rigid, shoemakers, for example, were prohibited from tannin hides.

The industrial revolution was facilitated by two elements: interchangeability of parts and division of labor. Division of labor has always been a feature from the beginning of civilization, the extent to which the division is carried out varied considerably depending on period and location. Compared to the Middle Ages, the Renaissance and the Age of Discovery were characterized by a greater specialization in labor, one of the characteristics of growing European cities and trade. It was in the late eighteenth century that Eli Whitney popularized the concept of interchangeability of parts when he manufactured 10,000 muskets. Up to this point in the history of manufacturing, each product (e.g. each gun) was considered a special order, meaning that parts of a given gun were fitted only for that particular gun and could not be used in other guns. Interchangeability of parts allowed the mass production of parts independent of the final products in which they would be used.

In 1883, Frederick Winslow Taylor introduced the stopwatch method for accurately measuring the time to perform each single task of a complicated job. He developed the scientific study of productivity and identifying how to coordinate different tasks to eliminate wasting of time and increase the quality of work. The next generation of scientific study occurred with the development of work sampling and predetermined motion time systems (PMTS). Work sampling is used to measure the random variable associated with the time of each task. PMTS allows the use of standard predetermined tables of the smallest body movements (e.g. turning the left wrist by 90°), and integrating them to predict the time needed to perform a simple task. PMTS has gained substantial importance due to the fact that it can predict work measurements without observing the actual work. The foundation of PMTS was laid out by the research and development of Frank B. and Lillian M. Gilbreth around 1912. The Gilbreths took advantage of taking motion pictures at known time intervals while operators were performing the given task.

The idea of the production line has been used multiple times in history prior to Henry Ford: the Venetian Arsenal (1104); Smith's pin manufacturing, in the Wealth of Nations (1776) or Brunel's Portsmouth Block Mills (1802). Ransom Olds was the first to manufacture cars using the assembly line system, but Henry Ford developed the first auto assembly system where a car chassis was moved through the assembly line by a conveyor belt while workers added components to it until the car was completed. During World War II, the growth of computing power led to further development of efficient manufacturing methods and the use of advanced mathematical and statistical tools. This was supported by the development of academic programs in industrial and systems engineering disciplines, as well as fields of operations research and management science (as multi-disciplinary fields of problem solving). While systems engineering concentrated on the broad characteristics of the relationships between inputs and outputs of generic systems, operations researchers concentrated on solving specific and focused problems. The synergy of operations research and systems engineering allowed for the realization of solving large scale and complex problems in the modern era. Recently, the development of faster and smaller computers, intelligent systems, and the World Wide Web has opened new opportunities for operations, manufacturing, production, and service systems.

Malakooti (2013) states that production and operation systems can be divided into five phases:

1. Empiricism (learning from experience)
2. Analysis (scientific management)

3. Synthesis (development of mathematical problem solving tools)
4. Isolated Systems with Single Objective (use of Integrated and Intelligent Systems, and WWW)
5. Integrated Complex Systems with Multiple Objectives (development of ecologically sound systems, environmentally sustainable systems, considering individual preferences)

Industrial Revolution

Marshall's flax mill in Holbeck. The textile industry is the prototypical example of the English industrial revolution.

Before the First industrial revolution work was mainly done through two systems: domestic system and craft guilds. In the domestic system merchants took materials to homes where artisans performed the necessary work, craft guilds on the other hand were associations of artisans which passed work from one shop to another, for example: leather was tanned by a tanner, passed to curriers, and finally arrived at shoemakers and saddlers.

The beginning of the industrial revolution is usually associated with 18th century English textile industry, with the invention of flying shuttle by John Kay in 1733, the spinning jenny by James Hargreaves in 1765, the water frame by Richard Arkwright in 1769 and the steam engine by James Watt in 1765. In 1851 at the Crystal Palace Exhibition the term American system of manufacturing was used to describe the new approach that was evolving in the United States of America which was based on two central features: interchangeable parts and extensive use of mechanization to produce them.

Henry Ford was 39 years old when he founded the Ford Motor Company in 1903, with $28,000 capital from twelve investors. The model T car was introduced in 1908, however it was not until Ford implemented the assembly line concept, that his vision of making a popular car affordable by every middle-class American citizen would be realized. The first factory in which Henry Ford used the concept of the assembly line was Highland Park (1913), he characterized the system as follows:

"The thing is to keep everything in motion and take the work to the man and not the man to the work. That is the real principle of our production, and conveyors are only one of many means to an end"

This became one the central ideas that led to mass production, one of the main elements of the Second Industrial Revolution, along with emergence of the electrical industry and petroleum industry.

Operations Management

Although productivity benefited considerably from technological inventions and division of labour, the problem of systematic measurement of performances and the calculation of these by the use of remained somewhat unexplored until Frederick Taylor, whose early work focused on developing what he called a "differential piece-rate system" and a series of experiments, measurements and formulas dealing with cutting metals and manual labor. The differential piece-rate system consisted in offering two different pay rates for doing a job: a higher rate for workers with high productivity (efficiency) and who produced high quality goods (effectiveness) and a lower rate for those who fail to achieve the standard. One of the problems Taylor believed could be solved with this system, was the problem of soldiering: faster workers reducing their production rate to that of the slowest worker. In 1911 Taylor published his "The Principles of Scientific Management", in which he characterized scientific management (also known as Taylorism) as:

1. The development of a true science;
2. The scientific selection of the worker;
3. The scientific education and development of the worker;
4. Intimate friendly cooperation between the management and the workers.

Taylor is also credited for developing stopwatch time study, this combined with Frank and Lillian Gilbreth motion study gave way to time and motion study which is centered on the concepts of standard method and standard time. Frank Gilbreth is also responsible for introducing the flow process chart in 1921. Other contemporaries of Taylor worth remembering are Morris Cooke (rural electrification in the 1920s and implementer of Taylor's principles of scientific management in the Philadelphia's Department of Public Works), Carl Barth (speed-and-feed-calculating slide rules) and Henry Gantt (Gantt chart). Also in 1910 Hugo Diemer published the first industrial engineering book: Factory Organization and Administration.

In 1913 Ford Whitman Harris published his "How Many parts to make at once" in which he presented the idea of the economic order quantity model. He described the problem as follows:

"Interest on capital tied up in wages, material and overhead sets a maximum limit to the quantity of parts which can be profitably manufactured at one time; "setup costs" on the job fix the minimum. Experience has shown one manager a way to determine the economical size of lots"

This paper inspired a large body of mathematical literature focusing on the problem of production planning and inventory control.

In 1924 Walter Shewhart introduced the control chart through a technical memorandum while working at Bell Labs, central to his method was the distinction between common cause and special cause of variation. In 1931 Shewhart published his Economic Control of Quality of Manufactured Product, the first systematic treatment of the subject of Statistical Process Control (SPC).

In the 1940s methods-time measurement (MTM) was developed by H.B. Maynard, JL Schwab and GJ Stegemerten. MTM was the first of a series of predetermined motion time systems, predetermined in the sense that estimates of time are not determined in loco but are derived from an industry standard. This was explained by its originators in a book they published in 1948 called "Method-Time Measurement".

Up to this point in history, optimization techniques were known for a very long time, from the simple methods employed by F.W.Harris to the more elaborate techniques of the calculus of variations developed by Euler in 1733 or the multipliers employed by Lagrange in 1811, and computers were slowly being developed, first as analog computers by Sir William Thomson (1872) and James Thomson (1876) moving to the eletro-mechanical computers of Konrad Zuse (1939 and 1941). During World War II however, the development of mathematical optimization went through a major boost with the development of the Colossus computer, the first electronic digital computer that was all programmable, and the possibility to computationally solve large linear programming problems, first by Kantorovich in 1939 working for the Soviet government and latter on in 1947 with the simplex method of Dantzig. These methods are known today as belonging to the field of operations research.

From this point on a curious development took place: while in the United States the possibility of applying the computer to business operations led to the development of management software architecture such as MRP and successive modifications, and ever more sophisticated optimization techniques and manufacturing simulation software, in post-war Japan a series of events at Toyota Motor led to the development of the Toyota Production System (TPS) and Lean Manufacturing.

In 1943, in Japan, Taiichi Ohno arrived at Toyota Motor company. Toyota evolved a unique manufacturing system centered on two complementary notions: just in time (produce only what is needed) and autonomation (automation with a human touch). Regarding JIT, Ohno was inspired by American supermarkets: workstations functioned like a supermarket shelf where the customer can get products they need, at the time they need and in the amount needed, the workstation (shelf) is then restocked. Autonomation was developed by Toyoda Sakichi in Toyoda Spinning and Weaving: an automatically activated loom that was also foolproof, that is automatically detected problems. In 1983 J.N Edwards published his "MRP and Kanban-American style" in which he described JIT goals in terms of seven zeros: zero defects, zero (excess) lot size, zero setups, zero breakdowns, zero handling, zero lead time and zero surging. This period also marks the spread of Total Quality Management (TQM) in Japan, ideas initially developed by American authors such as Deming, Juran and Armand V. Feigenbaum. TQM is a strategy for implementing and managing quality improvement on an organizational basis, this includes: participation, work culture, customer focus, supplier quality improvement and integration of the quality system with business goals. Schnonberger identified seven fundamentals principles essential to the Japanese approach:

1. Process control: SPC and worker responsibility over quality
2. Easy able -to-see quality: boards, gauges, meters, etc. and poka-yoke
3. Insistence on compliance: "quality first"
4. Line stop: stop the line to correct quality problems
5. Correcting one's own errors: worker fixed a defective part if he produced it
6. The 100% check: automated inspection techniques and foolproof machines
7. Continual improvement: ideally zero defects

In 1987 the International Organization for Standardization (ISO), recognizing the growing importance of quality, issued the ISO 9000, a family of standards related to quality management systems. There has been some controversy though regarding the proper procedures to follow and the amount of paperwork involved.

Meanwhile, in the sixties, a different approach was developed by George W. Plossl and Oliver W. Wight, this approach was continued by Joseph Orlicky as a response to the TOYOTA Manufacturing Program which led to Material Requirements Planning (MRP) at IBM, latter gaining momentum in 1972 when the American Production and Inventory Control Society launched the "MRP Crusade". One of the key insights of this management system was the distinction between dependent demand and independent demand. Independent demand is demand which originates outside of the production system, therefore not directly controllable, and dependent demand is demand for components of final products, therefore subject to being directly controllable by management through the bill of materials, via product design. Orlicky wrote "Materials Requirement Planning" in 1975, the first hard cover book on the subject. MRP II was developed by Gene Thomas at IBM, and expanded the original MRP software to include additional production functions. Enterprise resource planning (ERP) is the modern software architecture, which addresses, besides production operations, distribution, accounting, human resources and procurement.

Recent trends in the field revolve around concepts such as:

- Business Process Re-engineering (launched by Michael Hammer in 1993): a business management strategy focusing on the analysis and design of workflows and business processes within an organization. BPR seeks to help companies radically restructure their organizations by focusing on the ground-up design of their business processes.
- Lean Manufacturing: a systemic method for the elimination of waste ("Muda") within a manufacturing process. Lean also takes into account waste created through overburden ("Muri") and waste created through unevenness in work loads ("Mura"). The term lean manufacturing was coined in the book The Machine that Changed the World.
- Six Sigma (an approach to quality developed at Motorola between 1985-1987): Six Sigma refers to control limits placed at six (6) standard deviations from the mean of a normal distribution, this became very famous after Jack Welch of General Electric launched a company-wide initiative in 1995 to adopt this set of methods. More recently, Six Sigma has included DMAIC (for improving processes) and DFSS (for designing new products and new processes)
- Reconfigurable Manufacturing Systems: a production system designed at the outset for rapid change in its structure, as well as its hardware and software components, in order to quickly adjust its production capacity and functionality within a part family in response to sudden market changes or intrinsic system change.

Topics

Production Systems

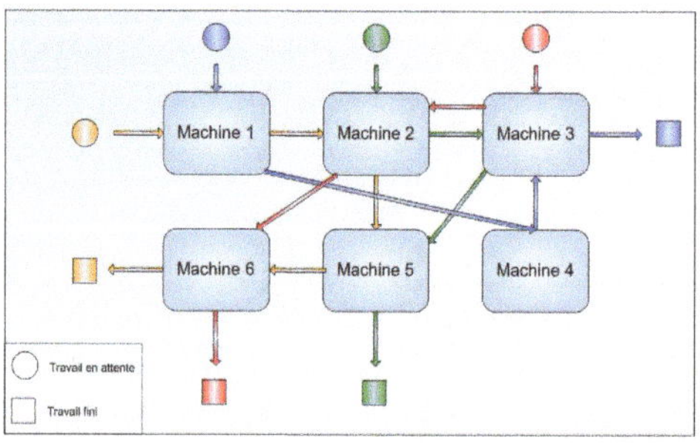

In a job shop machines are grouped by technological similarities regarding transformation processes, therefore a single shop can work very different products (in this picture four colors). Also notice that in this drawing each shop contains a single machine.

Flexible Manufacturing System: in the middle there are two rails for the shuttle to move pallets between machining centers (there are also FMS which use AGVs), in front of each machining center there is a buffer and in left we have a shelf for storing pallets. Usually in the back there is a similar system for managing the set of tools required for different machining operations.

A production system comprises both the technological elements (machines and tools) and organizational behavior (division of labor and information flow). An individual production system is usually analyzed in the literature referring to a single business, therefore it's usually improper to include in a given production system the operations necessary to process goods that are obtained by purchasing or the operations carried by the customer on the sold products, the reason being simply that since businesses need to design their own production systems this then becomes the focus of analysis, modeling and decision making (also called "configuring" a production system) .

A first possible distinction in production systems (technological classification) is between continuous process production and discrete part production (manufacturing).

- Process production means that the product undergoes physical-chemical transformations and lacks assembly operations, therefore the original raw materials can't easily be obtained from the final product, examples include: paper, cement, nylon and petroleum products.
- Part production (ex:cars and ovens) comprises both fabrication systems and assembly systems. In the first category we find job shops, manufacturing cells, flexible manufacturing systems and transfer lines, in the assembly category we have fixed position systems, assembly lines and assembly shops (both manual and/or automated operations).

Another possible classification is one based on Lead Time (manufacturing lead time vs delivery lead time): Engineer to Order (ETO, Purchase to Order (PTO), Make to Order (MTO), Assemble to Order (ATO) and Make to Stock (MTS). According to this classification different kinds of systems will have different customer order decoupling points (CODP), meaning that Work in Progress (WIP) cycle stock levels are practically nonexistent regarding operations located after the CODP (except for WIP due to queues).

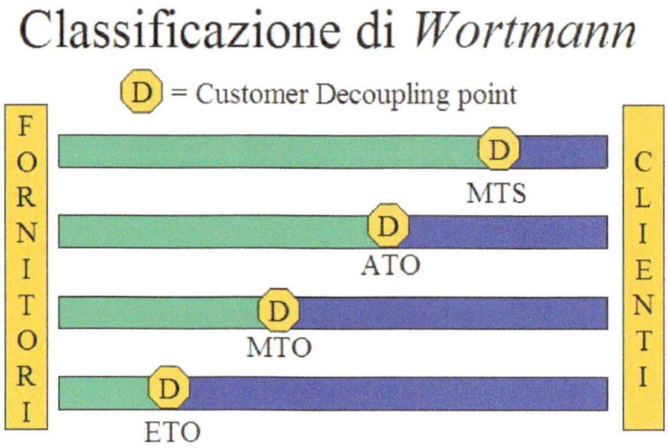

Delivery lead time is the blue bar, manufacturing time is the whole bar, the green bar is the difference between the two.

The concept of production systems can be expanded to the service sector world keeping in mind that services have some fundamental differences in respect to material goods: intangibility, client always present during transformation processes, no stocks for "finished goods". Services can be classified according to a service process matrix: degree of labor intensity (volume) vs degree of customization (variety). With a high degree of labor intensity there are Mass Services (e.g., commercial banking bill payments and state schools) and Professional Services (e.g., personal physicians and lawyers), while with a low degree of labor intensity there are Service Factories (e.g., airlines and hotels) and Service Shops (e.g., hospitals and auto mechanics).

The systems described above are ideal types: real systems may present themselves as hybrids of those categories. Consider, for example, that the production of jeans involves initially carding, spinning, dyeing and weaving, then cutting the fabric in different shapes and assembling the parts in pants or jackets by combining the fabric with thread, zippers and buttons, finally finishing and distressing the pants/jackets before being shipped to stores. The beginning can be seen as process production, the middle as part production and the end again as process production: it's unlikely that a single company will keep all the stages of production under a single roof, therefore the problem of vertical integration and outsourcing arises. Most products require, *from a supply chain perspective*, both process production and part production.

Metrics: Efficiency and Effectiveness

Operations strategy concerns policies and plans of use of the firm productive resources with the aim of supporting long term competitive strategy. Metrics in operations management can be broadly classified into efficiency metrics and effectiveness metrics. Effectiveness metrics involve:

1. Price (actually fixed by marketing, but lower bounded by production cost): purchase price, use costs, maintenance costs, upgrade costs, disposal costs
2. Quality: specification and compliance
3. Time: productive lead time, information lead time, punctuality
4. Flexibility: mix, volume, gamma
5. Stock availability
6. Ecological Soundness: biological and environmental impacts of the system under study.

A more recent approach, introduced by Terry Hill, involves distinguishing competitive variables in order winner and order qualifiers when defining operations strategy. Order winners are variables which permit differentiating the company from competitors, while order qualifiers are prerequisites for engaging in a transaction. This view can be seen as a unifying approach between operations management and marketing.

Productivity is a standard efficiency metric for evaluation of production systems, broadly speaking a ratio between outputs and inputs, and can assume many specific forms, for example: machine productivity, workforce productivity, raw material productivity, warehouse productivity (=inventory turnover). It is also useful to break up productivity in use U (productive percentage of total time) and yield η (ratio between produced volume and productive time) to better evaluate production systems performances. Cycle times can be modeled through manufacturing engineering if the individual operations are heavily automated, if the manual component is the prevalent one, methods used include: time and motion study, predetermined motion time systems and work sampling.

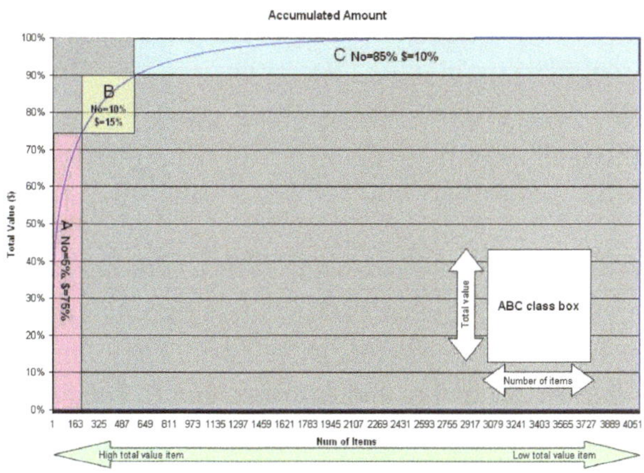

An ABC cumulated curve. Typically one curve is constructed for revenue (consumption) and another for inventory (stock).

ABC analysis is a method for analyzing inventory based on Pareto distribution, it posits that since revenue from items on inventory will be power law distributed then it makes sense to manage items differently based on their position on a revenue-inventory level matrix, 3 classes are constructed (A,B and C) from cumulative item revenues, so in a matrix each item will have a letter (A,B or C) assigned for revenue and inventory. This method posits that items away from the diagonal should be managed differently: items in the upper part are subject to risk of obsolescence, items in the lower part are subject to risk of stockout.

Throughput is a variable which quantifies the number of parts produced in the unit of time. Although estimating throughput for a single process maybe fairly simple, doing so for an entire production system involves an additional difficulty due to the presence of queues which can come from: machine breakdowns, processing time variability, scraps, setups, maintenance time, lack of orders, lack of materials, strikes, bad coordination between resources, mix variability, plus all these inefficiencies tend to compound depending on the nature of the production system. One important example of how system throughput is tied to system design are bottlenecks: in job shops bottlenecks are typically dynamic and dependent on scheduling while on transfer lines it makes sense to speak of "the bottleneck" since it can be univocally associated with a specific station on the line. This leads to the problem of how to define capacity measures, that is an estimation of the maximum output of a given production system, and capacity utilization.

Overall Equipment Effectiveness (OEE) is defined as the product between system availability, cycle time efficiency and quality rate. OEE is typically used as key performance indicator (KPI) in conjunction with the lean manufacturing approach.

Configuration and Management

Designing the *configuration of production systems* involves both technological and organizational variables. Choices in production technology involve: dimension-ing capacity, fractioning capacity, capacity location, outsourcing processes, process technology, automation of operations, trade-off between volume and variety. Choices in the organizational area involve: defining worker skills and responsibilities, team coordination, worker incentives and infor-mation flow.

Regarding *production planning*, there is a basic distinction between the push approach and the pull approach, with the later including the singular approach of just in time. Pull means that the production system authorizes production based on inventory level; push means that production occurs based on demand (forecasted or present, that is purchase orders). An individual production system can be both push and pull; for example activities before the CODP may work under a pull system, while activities after the CODP may work under a push system.

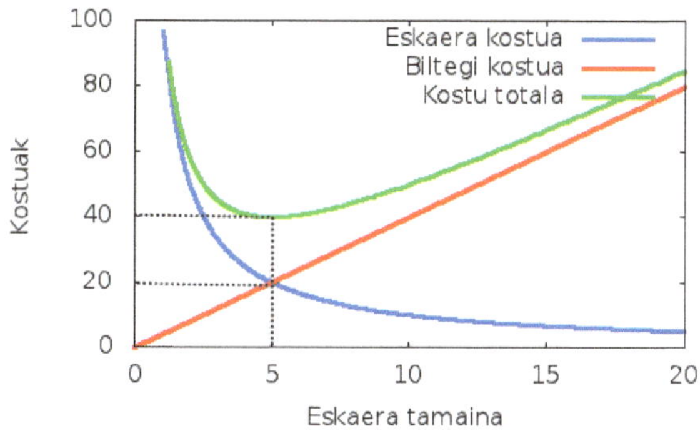

Classic EOQ model: trade-off between ordering cost (blue) and holding cost (red). Total cost (green) admits a global optimum.

Regarding the traditional pull approach to inventory control, a number of techniques have been developed based on the work of Ford W. Harris (1913), which came to be known as the economic order quantity (EOQ) model. This model marks the beginning of inventory theory, which includes the Wagner-Within Procedure, the News Vendor Model, Base Stock Model and the Fixed Time Period model. These models usually involve the calculation of cycle stocks and buffer stocks, the latter usually modeled as a function of demand variability. The Economic Production Quantity (EPQ) differs from the EOQ model only in that it assumes a constant fill rate for the part being produced, instead of the instantaneous refilling of the EOQ model.

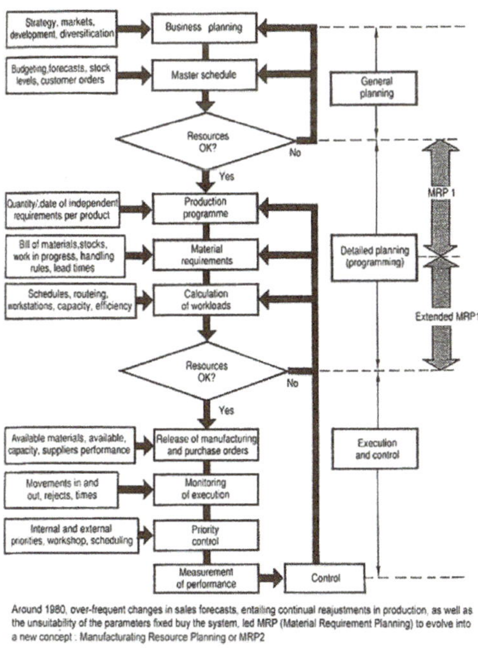

A typical MRPII construct: general planning (top) concerned with forecasts, capacity planning and inventory levels, programming (middle) concerned with calculation of workloads, rough-cut capacity planning, MPS, capacity requirements planning, traditional MRP planning, control (bottom) concerned with scheduling.

Joseph Orlickly and others at IBM developed a push approach to inventory control and production planning, now known as material requirements planning (MRP), which takes as input both the Master Production Schedule (MPS) and the Bill of Materials (BOM) and gives as output a schedule for the materials (components) needed in the production process. MRP therefore is a planning tool to manage purchase orders and production orders (also called jobs).

The MPS can be seen as a kind of aggregate planning for production coming in two fundamentally opposing varieties: plans which try to chase demand and level plans which try to keep uniform capacity utilization. Many models have been proposed to solve MPS problems:

- Analytical models (e.g. Magee Boodman model)
- Exact optimization algorithmic models (e.g. LP and ILP)
- Heuristic models (e.g. Aucamp model).

MRP can be briefly described as a 3s procedure: sum (different orders), split (in lots), shift (in time according to item lead time). To avoid an "explosion" of data processing in MRP (number of BOMs required in input) planning bills (such as family bills or super bills) can be useful since they allow a rationalization of input data into common codes. MRP had some notorious problems such as infinite capacity and fixed lead times, which influenced successive modifications of the original software architecture in the form of MRP II, enterprise resource planning (ERP) and Advanced planning and scheduling (APS).

In this context problems of scheduling (sequencing of production), loading (tools to use), part type selection (parts to work on) and applications of operations research have a significant role to play.

Lean manufacturing is an approach to production which arose in Toyota between the end of World War II and the seventies. It comes mainly from the ideas of Taiichi Ohno and Toyoda Sakichi which are centered on the complementary notions of just in time and autonomation (jidoka), all aimed at reducing waste (usually applied in PDCA style). Some additional elements are also fundamental: production smoothing (Heijunka), capacity buffers, setup reduction, cross-training and plant layout.

- Heijunka: production smoothing presupposes a level strategy for the MPS and a final assembly schedule developed from the MPS by smoothing aggregate production requirements in smaller time buckets and sequencing final assembly to achieve repetitive manufacturing. *If these conditions are met*, expected throughput can be equaled to the inverse of takt time. Besides volume, heijunka also means attaining mixed model production, which however may only be feasible through set-up reduction. A standard tool for achieving this is the Heijunka box.

- Capacity buffers: ideally a JIT system would work with zero breakdowns, this however is very hard to achieve in practice, nonetheless Toyota favors acquiring extra capacity over extra WIP to deal with starvation.
- Set-up reduction: typically necessary to achieve mixed model production, a key distinction can be made between internal and external setup. Internal setups (e.g. removing a die) refers to tasks when the machine is not working, while external setups can be completed while the machine is running (ex:transporting dies).
- Cross training: important as an element of Autonomation, Toyota cross trained their employees through rotation, this served as an element of production flexibility, holistic thinking and reducing boredom.
- Layout: U-shaped lines or cells are common in the lean approach since they allow for minimum walking, greater worker efficiency and flexible capacity.

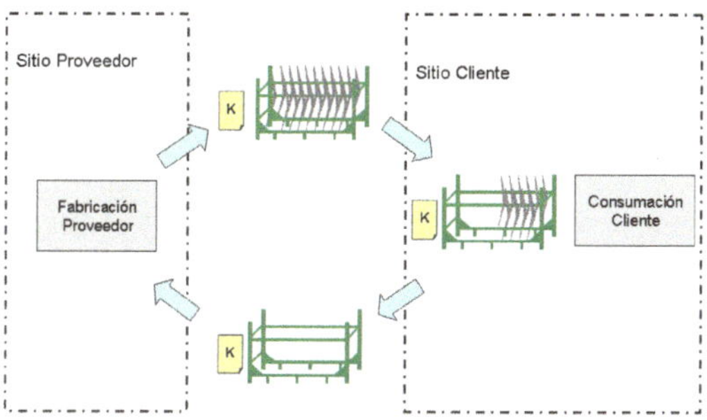

When introducing kanbans in real production systems, attaining unitary lot from the start maybe unfeasible, therefore the kanban will represent a given lot size defined by management.

A series of tools have been developed mainly with the objective of replicating Toyota success: a very common implementation involves small cards known as kanbans; these also come in some varieties: reorder kanbans, alarm kanbans, triangular kanbans, etc. In the classic kanban procedure with one card:

- Parts are kept in containers with their respective kanbans
- The downstream station moves the kanban to the upstream station and starts producing the part at the downstream station
- The upstream operator takes the most urgent kanban from his list (compare to queue discipline from queue theory) and produces it and attach its respective kanban

The two-card kanban procedure differs a bit:

- The downstream operator takes the production kanban from his list
- If required parts are available he removes the move kanban and places them in another box, otherwise he chooses another production card

- He produces the part and attach its respective production kanban
- Periodically a mover picks up the move kanbans in upstream stations and search for the respective parts, when found he exchanges production kanbans for move kanbans and move the parts to downstream stations

Since the number of kanbans in the production system is set by managers as a constant number, the kanban procedure works as WIP controlling device, which for a given arrival rate, per Little's Law, works as a lead time controlling device.

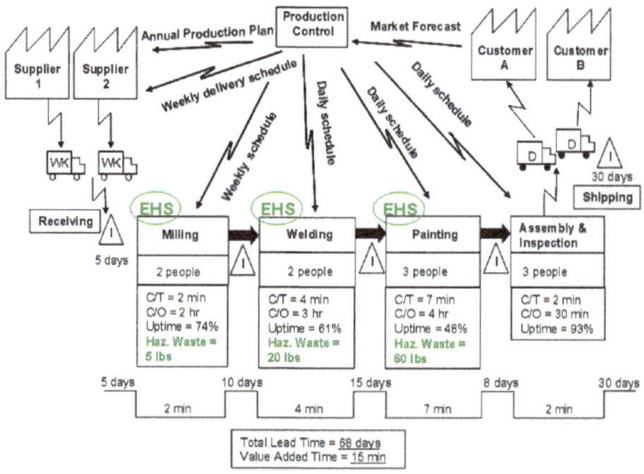

Value Stream Mapping, a representation of materials and information flows inside a company, mainly used in the lean manufacturing approach. The calculation of the time-line (bottom) usually involves using Little's Law to derive lead time from stock levels and takt time.

In Toyota the TPS represented more of a philosophy of production than a set of specific lean tools, the latter would include:

- SMED: a method for reducing changeover times
- Value Stream Mapping: a graphical method for analyzing the current state and designing a future state
- lot-size reduction
- elimination of time batching
- Rank Order Clustering: an algorithm which groups machines and product families together, used for designing manufacturing cells
- single-point scheduling, the opposite of the traditional push approach
- multi-process handling: when one operator is responsible for operating several machines or processes
- poka-yoke: any mechanism in lean manufacturing that helps an equipment operator avoid (*yokeru*) mistakes (*poka*)

- 5S: describes how to organize a work space for efficiency and effectiveness by identifying and storing the items used, maintaining the area and items, and sustaining the new order
- backflush accounting: a product costing approach in which costing is delayed until goods are finished

Seen more broadly, JIT can include methods such as: product standardization and modularity, group technology, total productive maintenance, job enlargement, job enrichment, flat organization and vendor rating (JIT production is very sensitive to replenishment conditions).

In heavily automated production systems production planning and information gathering may be executed via the control system, attention should be paid however to avoid problems suck as deadlocks, as these can lead to productivity losses.

Mathematical Modeling

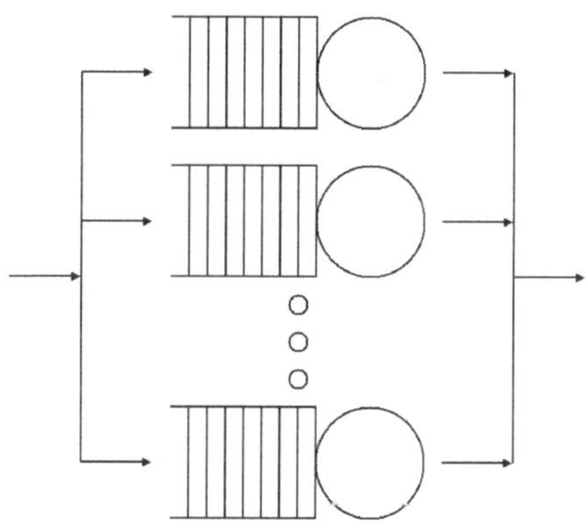

Queue networks are systems in which single queues are connected by a routing network. In this image servers are represented by circles, queues by a series of retangles and the routing network by arrows. In the study of queue networks one typically tries to obtain the equilibrium distribution of the network.

There are also fields of mathematical theory which have found applications in the field of operations management such as operations research: mainly mathematical optimization problems and queue theory. Queue theory is employed in modelling queue and processing times in production systems while mathematical optimization draws heavily from multivariate calculus and linear algebra. Queue theory is based on Markov chains and stochastic processes. It also worth noticing that computations of safety stocks are usually based on modeling demand as a normal distribution and MRP and some inventory problems can be formulated using optimal control.

When analytical models are not enough, managers may resort to using simulation. Simulation has been traditionally done thought the Discrete event simulation paradigm, where the simulation model possesses a state which can only change when a discrete event happens, which consists of a clock and list of events. The more recent Transaction-level modeling paradigm consists of a set of resources and a set of transactions: transactions move through a network of resources (nodes) according to a code, called process.

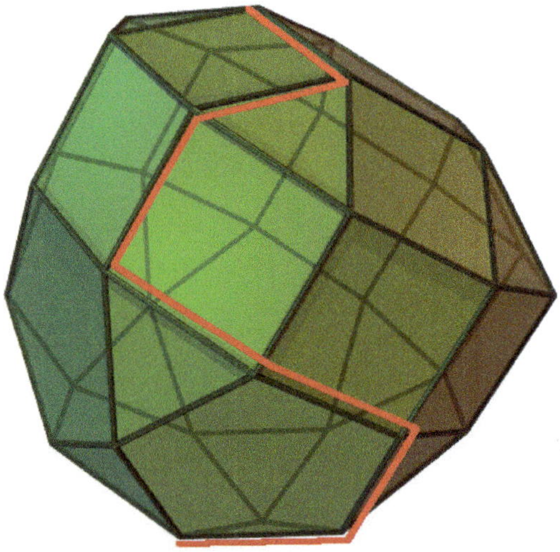

Illustration of the Simplex method, a classical approach to solving LP optimization problems and also integer programming (ex:branch and cut). Mainly used in push approach but also in production system configuration. The interior of the green polytope geometrically represents the feasible region, while the red line indicates the sequence of pivot operations required to reach the optimal solution.

Since real production processes are always affected by disturbances in both inputs and outputs, many companies implement some form of Quality management or quality control. The Seven Basic Tools of Quality designation provides a summary of commonly used tools:

- check sheets
- Pareto charts
- Ishikawa diagrams (Cause-and-effect diagram)
- control charts
- histogram
- scatter diagram
- stratification

These are used in approaches like Total quality management and Six Sigma. Keeping quality under control is relevant to both increasing customer satisfaction and reducing processing waste.

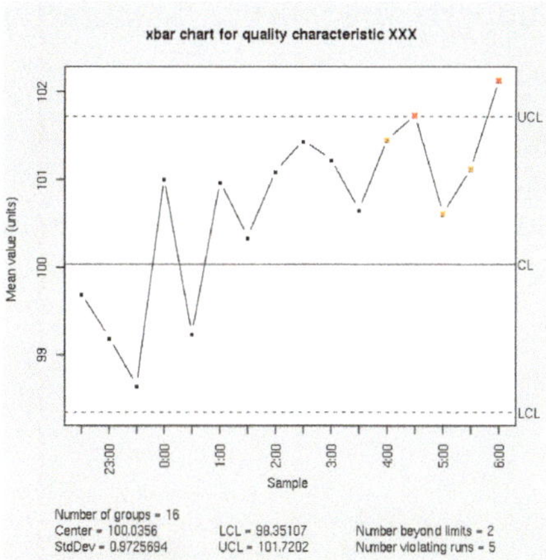

A control chart: process output variable is modeled by a probability density function and for each statistic of the sample an upper control line and lower control line are fixed, when the statistic moves out of bounds, an alarm is given and possible causes are investigated. In this drawing the statistic of choice is the mean and red points represent alarm points.

Operations management textbooks usually cover demand forecasting, even though it is not strictly speaking an operations problem, because demand is related to some production systems variables. For example, a classic approach in dimensioning safety stocks requires calculating standard deviation of forecast errors. Demand forecasting is also a critical part of push systems, since order releases have to be planned ahead of actual clients orders. Also any serious discussion of capacity planning involves adjusting company outputs with market demands.

Safety, Risk and Maintenance

Other important management problems involve maintenance policies, facility management and supply chain integration.

Organizations

The following organizations support and promote operations management:

- Association for Operations Management (APICS) which supports the *Production and Inventory Management Journal*
- European Operations Management Association (EurOMA) which supports the International Journal of Operations & Production Management
- Production and Operations Management Society (POMS) which supports the journal: *Production and Operations Management*
- Institute for Operations Research and the Management Sciences (INFORMS)
- The Manufacturing and Service Operations Management Society (MSOM) of

INFORMS which supports the journal: Manufacturing & Service Operations Management

- Institute of Operations Management (UK)
- Association of Technology, Management, and Applied Engineering (ATMAE)

Academic Job Market

Since January 2014 the following website has been created and dedicated to collecting and displaying valuable information related to the academic job market search of Operations Research/Management Science and related fields and disciplines:

- OperationsAcademia.org

The website operates as a multi-sided marketplace connecting job posters, job candidates and anyone interested in getting further insights on the academic job market in "Operations" (broadly defined). The website OperationsAcademia.org publicly shows available Job Postings, PhD Candidates on the market, and runs an Annual Job Market Survey. The interested visitor may also find information on recently hired Professors in Operations related departments, and all past job market postings and placements since 2014.

Management

Management (or managing) is the administration of an organization, whether it be a business, a not-for-profit organization, or government body. Management includes the activities of setting the strategy of an organization and coordinating the efforts of its employees or volunteers to accomplish its objectives through the application of available resources, such as financial, natural, technological, and human resources. The term "management" may also refer to the people who manage an organization.

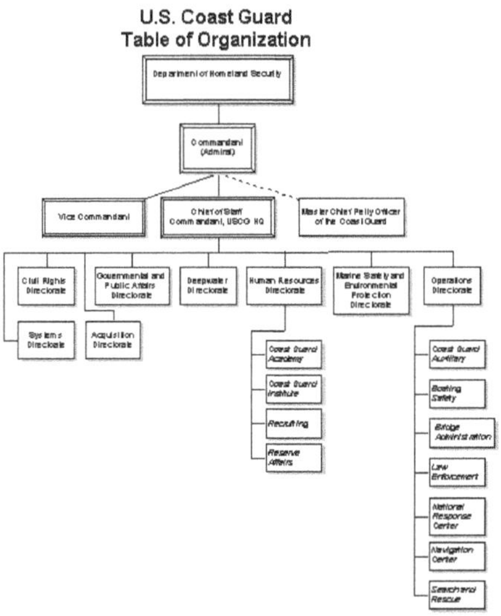

An organization chart for the United States Coast Guard shows the hierarchy of managerial roles in that organization.

Management is also an academic discipline, a social science whose objective is to study social organization and organizational leadership. Management is studied at colleges and universities; some important degrees in management are the Bachelor of Commerce (B.Com.) and Master of Business Administration (M.B.A.) and, for the public sector, the Master of Public Administration (MPA) degree. Individuals who aim at becoming management researchers or professors may complete the Doctor of Business Administration (DBA) or the PhD in business administration or management.

In larger organizations, there are generally three levels of managers, which are typically organized in a hierarchical, pyramid structure. Senior managers, such as the Board of Directors, Chief Executive Officer (CEO) or President of an organization, set the strategic goals of the organization and make decisions on how the overall organization will operate. Senior managers provide direction to the middle managers who report to them. Middle managers, examples of which would include branch managers, regional managers and section managers, provide direction to front-line managers. Middle managers communicate the strategic goals of senior management to the front-line managers. Lower managers, such as supervisors and front-line team leaders, oversee the work of regular employees (or volunteers, in some voluntary organizations) and provide direction on their work.

In smaller organizations, the roles of managers have much wider scopes. A manager can perform several roles or even all of the roles commonly observed in a large organization. There are many more smaller organizations than larger ones.

Definitions

Views on the definition and scope of management include:

- According to Henri Fayol, "to manage is to forecast and to plan, to organise, to command, to co-ordinate and to control."
- Fredmund Malik defines it as "the transformation of resources into utility."
- Management included as one of the factors of production - along with machines, materials and money.
- Ghislain Deslandes defines it as "a vulnerable force, under pressure to achieve results and endowed with the triple power of constraint, imitation and imagination, operating on subjective, interpersonal, institutional and environmental levels".
- Peter Drucker (1909–2005) saw the basic task of management as twofold: marketing and innovation. Nevertheless, innovation is also linked to marketing (product innovation is a central strategic marketing issue). Peter Drucker identifies marketing as a key essence for business success, but management and marketing are generally understood as two different branches of business administration knowledge.

Theoretical Scope

Management involves identifying the mission, objective, procedures, rules and manipulation of the human capital of an enterprise to contribute to the success of the enterprise. This implies effective communication: an enterprise environment (as opposed to a physical or mechanical mechanism) implies human motivation and implies some sort of successful progress or system outcome. As such, management is not the manipulation of a mechanism (machine or automated program), not the herding of animals, and can occur either in a legal or in an illegal enterprise or environment. Management does not need to be seen from enterprise point of view alone, because management is an essential function to improve one's life and relationships. Management is therefore everywhere and it has a wider range of application. Based on this, management must have humans, communication, and a positive enterprise endeavor. Plans, measurements, motivational psychological tools, goals, and economic measures (profit, etc.) may or may not be necessary components for there to be management. At first, one views management functionally, such as measuring quantity, adjusting plans, meeting goals. This applies even in situations where planning does not take place. From this perspective, Henri Fayol (1841–1925) considers management to consist of six functions:

1. forecasting
2. planning

3. organizing
4. commanding
5. coordinating
6. controlling

(Henri Fayol was one of the most influential contributors to modern concepts of management.)

In another way of thinking, Mary Parker Follett (1868–1933), allegedly defined management as "the art of getting things done through people". She described management as philosophy.

Critics, however, find this definition useful but far too narrow. The phrase "management is what managers do" occurs widely, suggesting the difficulty of defining management without circularity, the shifting nature of definitions and the connection of managerial practices with the existence of a managerial cadre or of a class.

One habit of thought regards management as equivalent to "business administration" and thus excludes management in places outside commerce, as for example in charities and in the public sector. More broadly, every organization must "manage" its work, people, processes, technology, etc. to maximize effectiveness. Nonetheless, many people refer to university departments that teach management as "business schools". Some such institutions (such as the Harvard Business School) use that name, while others (such as the Yale School of Management) employ the broader term "management".

English-speakers may also use the term "management" or "the management" as a collective word describing the managers of an organization, for cxamplc of a corporation. Historically this use of the term often contrasted with the term "labor" - referring to those being managed.

But in the present era the concept of management is identified in the wide areas and its frontiers have been pushed to a broader range. Apart from profitable organizations even non-profitable organizations (NGOs) apply management concepts. The concept and its uses are not constrained. Management on the whole is the process of planning, organizing, staffing, leading and controlling.

Nature of Work

In profitable organizations, management's primary function is the satisfaction of a range of stakeholders. This typically involves making a profit (for the shareholders), creating valued products at a reasonable cost (for customers), and providing great employment opportunities for employees. In nonprofit management, add the importance of keeping the faith of donors. In most models of management and governance, shareholders vote for the board of directors, and the board then hires senior management. Some organizations have experimented with other methods (such as employee-voting models) of selecting or reviewing managers, but this is rare.

In the public sector of countries constituted as representative democracies, voters elect politicians to public office. Such politicians hire many managers and administrators, and in some countries like the United States political appointees lose their jobs on the election of a new president/governor/mayor.

History

Some see management (by definition) as late-modern (in the sense of late modernity) conceptualization. On those terms it cannot have a pre-modern history, only harbingers (such as stewards). Others, however, detect management-like-thought back to Sumerian traders and to the builders of the pyramids of ancient Egypt. Slave-owners through the centuries faced the problems of exploiting/motivating a dependent but sometimes unenthusiastic or recalcitrant workforce, but many pre-industrial enterprises, given their small scale, did not feel compelled to face the issues of management systematically. However, innovations such as the spread of Hindu numerals (5th to 15th centuries) and the codification of double-entry book-keeping (1494) provided tools for management assessment, planning and control.

With the changing workplaces of industrial revolutions in the 18th and 19th centuries, military theory and practice contributed approaches to managing the newly-popular factories.

Given the scale of most commercial operations and the lack of mechanized record-keeping and recording before the industrial revolution, it made sense for most owners of enterprises in those times to carry out management functions by and for themselves. But with growing size and complexity of organizations, the split between owners (individuals, industrial dynasties or groups of shareholders) and day-to-day managers (independent specialists in planning and control) gradually became more common.

Etymology

The English verb "manage" comes from the Italian *maneggiare* (to handle, especially tools or a horse), which derives from the two Latin words *manus* (hand) and *agere* (to act). The French word for housekeeping, *ménagerie*, derived from *ménager* ("to keep house"; compare *ménage* for "household"), also encompasses taking care of domestic animals. *Ménagerie* is the French translation of Xenophon's famous book *Oeconomicus* on household matters and husbandry. The French word *mesnagement* (or *ménagement*) influenced the semantic development of the English word *management* in the 17th and 18th centuries.

Early Writing

While management (according to some definitions) has existed for millennia, several writers have created a background of works that assisted in modern management theories. Some ancient military texts have been cited for lessons that civilian managers can gather. For example, Chinese general Sun Tzu in the 6th century BC, *The Art of War*, recommends being aware of and acting on strengths and weaknesses of both a manager's organization and a foe's. The writings of influential Chinese Legalist philosopher Shen Buhai may be considered to embody a rare premodern example of abstract theory of administration.

Various ancient and medieval civilizations have produced "mirrors for princes" books, which aim to advise new monarchs on how to govern. Plato described job specialization in 350 B.C., and Alfarabi listed several leadership traits in A.D. 900. Other examples include the Indian Arthashastra by Chanakya (written around 300 BCE), and *The Prince* by Italian author Niccolò Machiavelli (c. 1515).

Written in 1776 by Adam Smith, a Scottish moral philosopher, *The Wealth of Nations* discussed efficient organization of work through division of labour. Smith described how changes in processes could boost productivity in the manufacture of pins. While individuals could produce 200 pins per day, Smith analyzed the steps involved in manufacture and, with 10 specialists, enabled production of 48,000 pins per day.

19th Century

Classical economists such as Adam Smith (1723–1790) and John Stuart Mill (1806–1873) provided a theoretical background to resource-allocation, production, and pricing issues. About the same time, innovators like Eli Whitney (1765–1825), James Watt (1736–1819), and Matthew Boulton (1728–1809) developed elements of technical production such as standardization, quality-control procedures, cost-accounting, interchangeability of parts, and work-planning. Many of these aspects of management existed in the pre-1861 slave-based sector of the US economy. That environment saw 4 million people, as the contemporary usages had it, "managed" in profitable quasi-mass production.

Salaried managers as an identifiable group first became prominent in the late 19th century.

20th Century

By about 1900 one finds managers trying to place their theories on what they regarded as a thoroughly scientific basis. Examples include Henry R. Towne's *Science of management* in the 1890s, Frederick Winslow Taylor's *The Principles of Scientific Management* (1911), Lillian Gilbreth's *Psychology of Management* (1914), Frank and Lillian Gilbreth's *Applied motion study* (1917) and Henry L. Gantt's charts (1910s). J. Duncan wrote the first college manage-ment-textbook in 1911. In 1912 Yoichi Ueno introduced Taylorism to Japan and became the first management consultant of the "Japanese-management style". His son Ichiro Ueno pioneered Japanese quality assurance.

The first comprehensive theories of management appeared around 1920. The Harvard Business School offered the first Master of Business Administration degree (MBA) in 1921. People like Henri Fayol (1841–1925) and Alexander Church described the various branches of management and their inter-relationships. In the early 20th century, people like Ordway Tead (1891–1973), Walter Scott and J. Mooney applied the principles of psychology to management. Other writers, such as Elton Mayo (1880–1949), Mary Parker Follett (1868–1933), Chester Barnard (1886–1961), Max Weber (1864–1920), who saw what he called the "administrator" as bureaucrat), Rensis Likert (1903–1981), and Chris Argyris (born 1923) approached the phenomenon of management from a sociological perspective.

Peter Drucker (1909–2005) wrote one of the earliest books on applied management: *Concept of the Corporation* (published in 1946). It resulted from Alfred Sloan (chairman of General Motors until 1956) commissioning a study of the organisation. Drucker went on to write 39 books, many in the same vein.

H. Dodge, Ronald Fisher (1890–1962), and Thornton C. Fry introduced statistical techniques into management-studies. In the 1940s, Patrick Blackett worked in the development of the applied-mathematics science of operations research, initially for military operations. Operations research, sometimes known as "management science" (but distinct from Taylor's scientific management), attempts to take a scientific approach to solving decision-problems, and can apply directly to multiple management problems, particularly in the areas of logistics and operations.

Some of the more recent developments include the Theory of Constraints, management by objectives, reengineering, Six Sigma and various information-technology-driven theories such as agile software development, as well as group-management theories such as Cog's Ladder.

As the general recognition of managers as a class solidified during the 20th century and gave perceived practitioners of the art/science of management a certain amount of prestige, so the way opened for popularised systems of management ideas to peddle their wares. In this context many management fads may have had more to do with pop psychology than with scientific theories of management.

Towards the end of the 20th century, business management came to consist of six separate branches, namely:

1. financial management
2. human resource management

3. information technology management (responsible for management information systems)

4. marketing management

5. operations management or production management

6. strategic management

21st Century

In the 21st century observers find it increasingly difficult to subdivide management into functional categories in this way. More and more processes simultaneously involve several categories. Instead, one tends to think in terms of the various processes, tasks, and objects subject to management.

Branches of management theory also exist relating to nonprofits and to government: such as public administration, public management, and educational management. Further, management programs related to civil-society organizations have also spawned programs in nonprofit management and social entrepreneurship.

Note that many of the assumptions made by management have come under attack from business-ethics viewpoints, critical management studies, and anti-corporate activism.

As one consequence, workplace democracy (sometimes referred to as Workers' self-management) has become both more common and advocated to a greater extent, in some places distributing all management functions among workers, each of whom takes on a portion of the work. However, these models predate any current political issue, and may occur more naturally than does a command hierarchy. All management embraces to some degree a democratic principle—in that in the long term, the majority of workers must support management. Otherwise, they leave to find other work or go on strike. Despite the move toward workplace democracy, command-and-control organization structures remain commonplace as *de facto* organization structure. Indeed, the entrenched nature of command-and-control is evident in the way that recent layoffs have been conducted with management ranks affected far less than employees at the lower levels. In some cases, management has even rewarded itself with bonuses after laying off lower-level workers.

According to leadership academic Manfred F.R. Kets de Vries, a contemporary senior management team will almost inevitably have some personality disorders.

Topics

Basics

Management operates through five basic functions: planning, organizing, coordinating, commanding, and controlling.

- Planning: Deciding what needs to happen in the future and generating plans for action(deciding in advance).
- Organizing: Making sure the human and nonhuman resources are put into place
- Coordinating (or staffing): Creating a structure through which an organization's goals can be accomplished.
- Commanding (or leading): Determining what must be done in a situation and getting people to do it.
- Controlling: Checking progress against plans.

Basic Roles

- Interpersonal: roles that involve coordination and interaction with employees
- Informational: roles that involve handling, sharing, and analyzing information
- Decision: roles that require decision-making

Skills

Management skills include:

- political: used to build a power base and to establish connections
- conceptual: used to analyze complex situations
- interpersonal: used to communicate, motivate, mentor and delegate
- diagnostic: ability to visualize appropriate responses to a situation
- leadership: ability to lead and to provide guidance to a specific group
- technical: expertise in one's particular functional area.

Implementation of Policies and Strategies

- All policies and strategies must be discussed with all managerial personnel and staff.
- Managers must understand where and how they can implement their policies and strategies.
- A plan of action must be devised for each department.
- Policies and strategies must be reviewed regularly.

- Contingency plans must be devised in case the environment changes.
- Top-level managers should carry out regular progress assessments.
- The business requires team spirit and a good environment.
- The missions, objectives, strengths and weaknesses of each department must be analyzed to determine their roles in achieving the business's mission.
- The forecasting method develops a reliable picture of the business's future environment.
- A planning unit must be created to ensure that all plans are consistent and that policies and strategies are aimed at achieving the same mission and objectives.

Policies and Strategies in the Planning Process

- They give mid and lower-level managers a good idea of the future plans for each department in an organization.
- A framework is created whereby plans and decisions are made.
- Mid and lower-level management may add their own plans to the business's strategies.

Levels

Most organizations have three management levels: first-level, middle-level, and top-level managers. First-line managers are the lowest level of management and manage the work of nonmanagerial individuals who are directly involved with the production or creation of the organization's products. First-line managers are often called supervisors, but may also be called line managers, office managers, or even foremen. Middle managers include all levels of management between the first-line level and the top level of the organization. These managers manage the work of first-line managers and may have titles such as department head, project leader, plant manager, or division manager. Top managers are responsible for making organization-wide decisions and establishing the plans and goals that affect the entire organization. These individuals typically have titles such as executive vice president, president, managing director, chief operating officer, chief executive officer, or chairman of the board.

These managers are classified in a hierarchy of authority, and perform different tasks. In many organizations, the number of managers in every level resembles a pyramid. Each level is explained below in specifications of their different responsibilities and likely job titles.

Top

The top or senior layer of management consists of the board of directors (including non-executive directors and executive directors), president, vice-president, CEOs and other members of the C-level executives. Different organizations have various members in their C-suite, which may include a Chief Financial Officer, Chief Technology Officer, and so on. They are responsible for controlling and overseeing the operations of the entire organization. They set a "tone at the top" and develop strategic plans, company policies, and make decisions on the overall direction of the organization. In addition, top-level managers play a significant role in the mobilization of outside resources. Senior managers are accountable to the shareholders, the general public and to public bodies that oversee corporations and similar organizations. Some members of the senior management may serve as the public face of the organization, and they may make speeches to introduce new strategies or appear in marketing.

The board of directors is typically primarily composed of non-executives who owe a fiduciary duty to shareholders and are not closely involved in the day-to-day activities of the organization, although this varies depending on the type (e.g., public versus private), size and culture of the organization. These directors are theoretically liable for breaches of that duty and typically insured under directors and officers liability insurance. Fortune 500 directors are estimated to spend 4.4 hours per week on board duties, and median compensation was $212,512 in 2010. The board sets corporate strategy, makes major decisions such as major acquisitions, and hires, evaluates, and fires the top-level manager (Chief Executive Officer or CEO). The CEO typically hires other positions. However, board involvement in the hiring of other positions such as the Chief Financial Officer (CFO) has increased. In 2013, a survey of over 160 CEOs and directors of public and private companies found that the top weaknesses of CEOs were "mentoring skills" and "board engagement", and 10% of companies never evaluated the CEO. The board may also have certain employees (e.g., internal auditors) report to them or directly hire independent contractors; for example, the board (through the audit committee) typically selects the auditor.

Helpful skills of top management vary by the type of organization but typically include a broad understanding of competition, world economies, and politics. In addition, the CEO is responsible for implementing and determining (within the board's framework) the broad policies of the organization. Executive management accomplishes the day-to-day details, including: instructions for preparation of department budgets, procedures, schedules; appointment of middle level executives such as department managers; coordination of departments; media and governmental relations; and shareholder communication.

Middle

Consist of general managers, branch managers and department managers. They are accountable to the top management for their department's function. They devote more time to organizational and directional functions. Their roles can be emphasized as executing organizational plans in conformance with the company's policies and the objectives of the top management, they define and discuss information and policies from top management to lower management, and most importantly they inspire and provide guidance to lower level managers towards better performance.

Middle management is the midway management of a categorized organization, being secondary to the senior management but above the deepest levels of operational members. An operational manager may be well-thought-out by middle management, or may be categorized as non-management operate, liable to the policy of the specific organization. Efficiency of the middle level is vital in any organization, since they bridge the gap between top level and bottom level staffs.

Their functions include:

- Design and implement effective group and inter-group work and information systems.
- Define and monitor group-level performance indicators.
- Diagnose and resolve problems within and among work groups.
- Design and implement reward systems that support cooperative behavior. They also make decision and share ideas with top managers.

Lower

Lower managers include supervisors, section leaders, forepersons and team leaders. They focus on controlling and directing regular employees. They are usually responsible for assigning employees' tasks, guiding and supervising employees on day-to-day activities, ensuring the quality and quantity of production and/or service, making recommendations and suggestions to employees on their work, and channeling employee concerns that they cannot resolve to mid-level managers or other administrators. First-level or "front line" managers also act as role models for their employees. In some types of work, front line managers may also do some of the same tasks that employees do, at least some of the time. For example, in some restaurants, the front line managers will also serve customers during a very busy period of the day.

Front-line managers typically provide:

- Training for new employees
- Basic supervision
- Motivation
- Performance feedback and guidance

Some front-line managers may also provide career planning for employees who aim to rise within the organization.

Training

Colleges and universities around the world offer bachelor's degrees, graduate degrees, diplomas and certificates in management, generally within their colleges of business, business schools or faculty of management but also in other related departments. In the 2010s, there has been an increase in online management education and training in the form of electronic educational technology (also called e-learning). Online education has increased the accessibility of management training to people who do not live near a college or university, or who cannot afford to travel to a city where such training is available.

While some professions require academic credentials in order to work in the profession (e.g., law, medicine, engineering, which require, respectively the Bachelor of Law, Doctor of Medicine and Bachelor of Engineering degrees), management and administration positions do not necessarily require the completion of academic degrees. Some well-known senior executive in the United States who did not complete a university degree include Steve Jobs, Bill Gates and Mark Zuckerberg. However, many managers and executives have completed some type of business or management training, such as a Bachelor of Commerce or a Master of Business Administration degree. Some major organizations, including companies, not-for-profit organizations and governments, require applicants to managerial or executive positions to hold at minimum Bachelor's degree in a field related to administration of management, or in the case of business jobs, a Bachelor of Commerce or a similar degree.

United States

Undergraduate

At the undergraduate level, the most common business program is the Bachelor of Commerce (B.Com.). A B.Com. is typically a four-year program that includes courses that give students an overview of the role of managers in planning and directing within an organization. Course topics include accounting, financial management, statistics, marketing, strategy, and other related areas. There are many other undergraduate degrees that include the study of management, such as Bachelor of Arts degrees with a major in business administration or management and Bachelor of Public Administration (B.P.A), a degree designed for individuals aiming to work as bureaucrats in the government jobs. Many colleges and universities also offer certificates and diplomas in business administration or management, which typically require one to two years of full-time study.

Graduate

At the graduate level students aiming at careers as managers or executives may choose to specialize in major subareas of management or business administration such as entrepreneurship, human resources, international business, organizational behavior, organizational theory, strategic management, accounting, corporate finance, entertainment, global management, healthcare management, investment management, sustainability and real estate. A Master of Business Administration (MBA) is the most popular professional master's degree and can be obtained from many universities in the United States. MBAs provide further education in management and leadership for graduate students. Other master's degrees in business and management include the Master of Science (M.Sc.) in business administration or management, which is typically taken by students aiming to become researchers or professors. There are also specialized master's degrees in administration for individuals aiming at careers outside of business, such as the Master of Public Administration (MPA) degree (also offered as a Master of Arts in Public Administration in some universities), for students aiming to become managers or executives in the public service and the Master of Health Administration, for students aiming to become managers or executives in the health care and hospital sector.

Management doctorates are the most advanced terminal degrees in the field of business and management. Most individuals obtaining management doctorates take the programs to obtain the training in research methods, statistical analysis and writing academic papers that they will need to seek careers as researchers, senior consultants and/or professors in business administration or management. There are two main types of management doctorates: the Doctor of Business Administration (D.B.A.) and the PhD in business administration or management. In the 2010s, doctorates in business administration and management are available with many specializations.

Good Practices

While management trends can change rapidly, the long term trend in management has been defined by a market embracing diversity and a rising service industry. Managers are currently being trained to encourage greater equality for minorities and women in the workplace, by offering increased flexibility in working hours, better retraining, and innovative (and usually industry-specific) performance markers. Managers destined for the service sector are being trained to use unique measurement techniques, better worker support and more charismatic leadership styles. Human resources finds itself increasingly working with management in a training capacity to help collect management data on the success (or failure) of management actions with employees.

References

- SS Gulshan. Management Principles and Practices by Lallan Prasad and SS Gulshan. Excel Books India. pp. 6–. ISBN 978-93-5062-099-1.
- Malakooti, Behnam (2013). Operations and Production Systems with Multiple Objectives. John Wiley & Sons. ISBN 978-1-118-58537-5.

- OperationsAcademia.org: showing the state-of-the-art of research in Operations Research/Management Science and related disciplines Retrieved on October 22, 2016
- "The Role of HR in Uncertain Times" (PDF). Economist Intelligence Unit. Economist Intelligence Unit. Retrieved 18 January 2015.
- "1st International Workshop on Business Process Management in the Cloud". KIT. 26 August 2013. Retrieved 13 March 2013.
- Keith Harrison-Broninski (2005-04-03). "Human Interaction: The Missing Link in BPM (Part I)". Retrieved 2013-08-16.

2

Business Model and its Types

Business models are models that are used as abstracts to represent an organization. It is used to describe the procedure of how an organization is created and delivered. The process of contrasting this model is a part of business strategy. The topics discussed in the chapter are of great importance to broaden the existing knowledge on business models.

Business Model

A business model is an "abstract representation of an organization, be it conceptual, textual, and/or graphical, of all core interrelated architectural, co-operational, and financial arrangements designed and developed by an organization presently and in the future, as well as all core products and/or services the organization offers, or will offer, based on these arrangements that are needed to achieve its strategic goals and objectives." This definition by Al-Debei, El-Haddadeh and Avison (2008) indicates that value proposition, value architecture (the organizational infrastructure and technological architecture that allows the movement of products, services, and information), value finance (modeling information related to total cost of ownership, pricing methods, and revenue structure), and value network articulate the primary constructs or dimensions of business models.

A business model describes the rationale of how an organization creates, delivers, and captures value, in economic, social, cultural or other contexts. The process of business model construction is part of business strategy.

In theory and practice, the term *business model* is used for a broad range of informal and formal descriptions to represent core aspects of a business, including purpose, business process, target customers, offerings, strategies, infrastructure, organizational structures, sourcing, trading practices, and operational processes and policies including culture. The literature has provided very diverse interpretations and definitions of a business model. A systematic review and analysis of manager responses to a survey defines business models as the design of organizational structures to enact a commercial opportunity. Further extensions to this design logic emphasize the use of narrative or coherence in business model descriptions as mechanisms by which entrepreneurs create extraordinarily successful growth firms.

Business models are used to describe and classify businesses, especially in an entrepreneurial setting, but they are also used by managers inside companies to explore possibilities for future development. Well-known business models can operate as "recipes" for creative managers. Business models are also referred to in some instances within the context of accounting for purposes of public reporting.

History

Over the years, business models have become much more sophisticated. The *bait and hook* business model (also referred to as the "razor and blades business model" or the "tied products business model") was introduced in the early 20th century. This involves offering a basic product at a very low cost, often at a loss (the "bait"), then charging compensatory recurring amounts for refills or associated products or services (the "hook"). Examples include: razor (bait) and blades (hook); cell phones (bait) and air time (hook); computer printers (bait) and ink cartridge refills (hook); and cameras (bait) and prints (hook). A variant of this model is Adobe, a software developer that gives away its document reader free of charge but charges several hundred dollars for its document writer.

In the 1953s, new business models came from McDonald's Restaurants and Toyota. In the 1960s, the innovators were Wal-Mart and Hypermarkets. The 1970s saw new business models from FedEx and Toys R Us; the 1980s from Blockbuster, Home Depot, Intel, and Dell Computer; the 1990s from Southwest Airlines, Netflix, eBay, Amazon.com, and Starbucks.

Today, the type of business models might depend on how technology is used. For example, entrepreneurs on the internet have also created entirely new models that depend entirely on existing or emergent technology. Using technology, businesses can reach a large number of customers with minimal costs. In addition, the rise of outsourcing and globalization has meant that business models must also account for strategic sourcing, complex supply chains and moves to collaborative, relational contracting structures.

Theoretical and Empirical Insights

Design Logic and Narrative Coherence

Design logic views the business model as an outcome of creating new organizational structures or changing existing structures to pursue a new opportunity. Gerry George and Adam Bock (2011) conducted a comprehensive literature review and surveyed managers to understand how they perceived the components of a business model. In that analysis these authors show that there is a design logic behind how entrepreneurs and managers perceive and explain their business model. In further extensions to the design logic, George and Bock (2012) use case studies and the IBM survey data on business models in large companies, to describe how CEOs and entrepreneurs create narratives or stories in a coherent manner to move the business from one opportunity to another. They also show that when the narrative is incoherent or the components of the story are misaligned, that these businesses tend to fail. They recommend ways in which the entrepreneur or CEO can create strong narratives for change.

Complementarities of between Partnering Firms

Studying collaborative research and the accessing of external sources of technology, Hummel et al. (2010) found that in deciding on business partners, it is important to make sure that both parties' business models are complementary. For example, they found that it was important to identify the value drivers of potential partners by analyzing their business models, and that it is beneficial to find partner firms that understand key aspects of our own firm's business model.

The University of Tennessee conducted research into highly collaborative business relationships. Researchers codified their research into a sourcing business model known as Vested (also referred to as Vested Outsourcing). Vested is a hybrid sourcing business model in which buyers and suppliers in an outsourcing or business relationship focus on shared values and goals to create an arrangement that is highly collaborative and mutually beneficial to each.

Categorization

From about 2012, some research and experimentation has theorized about a so-called "liquid business model".

V^4 BM Framework

Al-Debei and Avison (2010) V^4 BM Framework - four main dimensions encapsulating sixteen elements: Value Proposition, Value Architecture, Value Network, and Value Finance

- Value Proposition: This dimension implies that a BM should include a description of the products/services a digital organization offers, or will offer, along with their related information. Furthermore, the BM needs also to describe the value elements incorporated within the offering, as well as the nature of targeted market segment(s) along with their preferences.
- Value Architecture: portrays the concept as a holistic structural design of an organization, including its technological architecture, organizational infrastructure, and their configurations.
- Value Network: depicts the cross-company or inter-organization perspective towards the concept and has gained much attention in the BM literature.
- Value Finance: depicts information related to costing, pricing methods, and revenue structure

Shift from Pipes to Platforms

Sangeet Paul Choudary (2013) distinguishes between two broad families of business models in an article in *Wired* magazine. Choudary contrasts pipes (linear business models) with platforms (networked business models). In the case of pipes, firms create goods and services, push them out and sell them to customers. Value is produced upstream and consumed downstream. There is a linear flow, much like water flowing through a pipe. Unlike pipes, platforms do not just create and push stuff out. They allow users to create and consume value.

In an op-ed on MarketWatch, Choudary, Van Alstyne and Parker further explain how business models are moving from pipes to platforms, leading to disruption of entire industries.

Platform

There are three elements to a successful platform business model. The *Toolbox* creates connection by making it easy for others to plug into the platform. This infrastructure enables interactions between participants. The *Magnet* creates pull that attracts participants to the platform. For transaction platforms, both producers and consumers must be present to achieve critical mass. The *Matchmaker* fosters the flow of value by making connections between producers and consumers. Data is at the heart of successful matchmaking, and distinguishes platforms from other business models.

Chen (2009) stated that the business model has to take into account the capabilities of Web 2.0, such as collective intelligence, network effects, user-generated content, and the possibility of self-improving systems. He suggested that the service industry such as the airline, traffic, transportation, hotel, restaurant, information and communications technology and online gaming industries will be able to benefit in adopting business models that take into account the characteristics of Web 2.0. He also emphasized that Business Model 2.0 has to take into account not just the technology effect of Web 2.0 but also the networking effect. He gave the example of the success story of Amazon in making huge revenues each year by developing an open platform that supports a community of companies that re-use Amazon's on-demand commerce services.

Applications

Malone et al. found that some business models, as defined by them, indeed performed better than others in a dataset consisting of the largest U.S. firms, in the period 1998 through 2002, while they did not prove whether the existence of a business model mattered.

In the context of the Software-Cluster, which is funded by the German Federal Ministry of Education and Research, a business model wizard for software companies has been developed. It supports the design and analysis of software business models. The tool's underlying concept and data were published in various scientific publications.

The concept of a business model has been incorporated into certain accounting standards. For example, the International Accounting Standards Board (IASB) utilizes an "entity's business model for managing the financial assets" as a criterion for determining whether such assets should be measured at amortized cost or at fair value in its financial instruments accounting standard, IFRS 9. In their 2013 proposal for accounting for financial instruments, the Financial Accounting Standards Board also proposed a similar use of business model for classifying financial instruments. The concept of business model has also been introduced into the accounting of deferred taxes under International Financial Reporting Standards with 2010 amendments to IAS 12 addressing deferred taxes related to investment property.

Both IASB and FASB have proposed using the concept of business model in the context of reporting a lessor's lease income and lease expense within their joint project on accounting for leases. In its 2016 lease accounting model, IFRS 16, the IASB chose not to include a criterion of "stand alone utility" in its lease definition because "entities might reach different conclusions for contracts that contain the same rights of use, depending on differences between customers' resources or suppliers' business models." The concept has also been proposed as an approach for determining the measurement and classification when accounting for insurance contracts. As a result of the increasing prominence the concept of business model has received in the context of financial reporting, the European Financial Reporting Advisory Group (EFRAG), which advises the European Union on endorsement of financial reporting standards, commenced a project on the "Role of the Business Model in Financial Reporting" in 2011.

Design

Business model design refers to the activity of designing a company's business model. It is part of the business development and business strategy process and involves design methods.

Definitions

Al-Debei and Avison (2010) define a business model as an abstract representation of an organization. This may be conceptual, textual, and/or graphical, of all core interrelated architectural, co-operational, and financial arrangements designed and developed by an organization presently and in the future, as well all core products and/or services the organization offers, or will offer, based on these arrangements that are needed to achieve its strategic goals and objectives. This definition indicates that value proposition, value architecture, value finance, and value network articulate the primary constructs or dimensions of business models.

Economic Consideration

Al-Debei and Avison (2010) consider value finance as one of the main dimensions of

BM which depicts information related to costing, pricing methods, and revenue structure. Stewart and Zhao (2000) defined the business model as *a statement of how a firm will make money and sustain its profit stream over time.*

Component Consideration

Osterwalder et al. (2005) consider the Business Model as the blueprint of how a company does business. Slywotzky (1996) regards the business model as *the totality of how a company selects its customers, defines and differentiates it offerings, defines the tasks it will perform itself and those it will outsource, configures its resources, goes to market, creates utility for customers and captures profits.*

Strategic Outcome

Mayo and Brown (1999) considered the business model as *the design of key interdependent systems that create and sustain a competitive business.* Casadesus-Masanell and Ricart (2011) explain a business model as a set of *choices (policy, assets and governance)* and *consequences (flexible and rigid)* and underline the importance of considering *how it interacts with models of other players in the industry* instead of thinking of it in isolation.

Definitions of Design or Development

Zott and Amit (2009) consider business model design from the perspectives of design themes and design content. Design themes refer to the system's dominant value creation drivers and design content examines in greater detail the activities to be performed, the linking and sequencing of the activities and who will perform the activities.

Design Themes Emphasis

Environment-Strategy-Structure-Operations (ESSO) Business Model Development

Developing a Framework for Business Model Development with an emphasis on Design Themes, Lim (2010) proposed the Environment-Strategy-Structure-Operations (ESSO) Business Model Development which takes into consideration the alignment of the organization's strategy with the organization's structure, operations, and the environmental factors in achieving competitive advantage in varying combination of cost, quality, time, flexibility, innovation and affective.

Design Content Emphasis

Business model design includes the modeling and description of a company's:

- value propositions

- target customer segments
- distribution channels
- customer relationships
- value configurations
- core capabilities
- commercial network
- partner network
- cost structure
- revenue model

Business model design is distinct from business modeling. The former refers to defining the business logic of a company at the strategic level, whereas the latter refers to business process design at the operational level.

A business model design template can facilitate the process of designing and describing a company's business model.

Daas et al. (2012) developed a decision support system (DSS) for business model design. In their study a decision support system (DSS) is developed to help SaaS in this process, based on a design approach consisting of a design process that is guided by various design methods.

Examples

In the early history of business models it was very typical to define business model types such as bricks-and-mortar or e-broker. However, these types usually describe only one aspect of the business (most often the revenue model). Therefore, more recent literature on business models concentrate on describing a business model as a whole, instead of only the most visible aspects.

The following examples provide an overview for various business model types that have been in discussion since the invention of term *business model*:

- Bricks and clicks business model

  Business model by which a company integrates both offline (*bricks*) and online (*clicks*) presences. One example of the bricks-and-clicks model is when a chain of stores allows the user to order products online, but lets them pick up their order at a local store.

- Collective business models

  Business system, organization or association typically composed of relatively large numbers of businesses, tradespersons or professionals in the same or related fields of endeavor, which pools resources, shares information or provides other benefits for their members. For example, a science park or high-tech campus provides shared resources (e.g. cleanrooms and other lab facilities) to the firms located on its premises, and in addition seeks to create an innovation community among these firms and their employees.

- Cutting out the middleman model

  The removal of intermediaries in a supply chain: "cutting out the middleman". Instead of going through traditional distribution channels, which had some type of intermediate (such as a distributor, wholesaler, broker, or agent), companies may now deal with every customer directly, for example via the Internet.

- Direct sales model

  Direct selling is marketing and selling products to consumers directly, away from a fixed retail location. Sales are typically made through party plan, one-to-one demonstrations, and other personal contact arrangements. A text book definition is: "The direct personal presentation, demonstration, and sale of products and services to consumers, usually in their homes or at their jobs."

- Distribution business models, various

- Fee in, free out

  Business model which works by charging the first client a fee for a service, while offering that service free of charge to subsequent clients.

- Franchise

  Franchising is the practice of using another firm's successful business model. For the franchisor, the franchise is an alternative to building 'chain stores' to *distribute* goods and avoid investment and liability over a chain. The franchisor's success is the success of the franchisees. The franchisee is said to have a greater incentive than a direct employee because he or she has a direct stake in the business.

- Sourcing business model

  A sourcing business model is a type of business model that is applied specifically to business relationships where more than one party needs to work with another party to be successful. It is the combination of two concepts: the contractual relationship framework a company uses with its supplier (transactional, relational, investment based), and the economic model used (transactional, output or outcome-based).

- Freemium business model

  Business model that works by offering basic Web services, or a basic downloadable digital product, for free, while charging a premium for advanced or special features.

- Pay what you can (PWYC)

  A non-profit or for-profit business model which does not depend on set prices for its goods, but instead asks customers to pay what they feel the product or service is worth to them. It is often used as a promotional tactic, but can also be the regular method of doing business. It is a variation on the gift economy and cross-subsidization, in that it depends on reciprocity and trust to succeed.

  "Pay what you want" (PWYW) is sometimes used synonymously, but "pay what you can" is often more oriented to charity or socially oriented uses, based more on *ability* to pay, while "pay what you want" is often more broadly oriented to perceived value in combination with willingness and ability to pay.

- Value-added reseller model

  Value Added Reseller is a model where a business makes something which is resold by other businesses but with modifications which add value to the original product or service. These modifications or additions are mostly industry specific in nature and are essential for the distribution. Businesses going for a VAR model have to develop a VAR network. It is one of the latest collaborative business models which can help in faster development cycles and is adopted by many Technology companies especially software.

Other examples of business models are:

- Auction business model
- All-in-one business model
- Chemical leasing
- Low-cost carrier business model
- Loyalty business models
- Monopolistic business model
- Multi-level marketing business model

- Network effects business model
- Online auction business model
- Online content business model
- Online media cooperative
- Premium business model
- Professional open-source model
- Pyramid scheme business model
- Razor and blades business model
- Servitization of products business model
- Subscription business model

Frameworks

Technology centric communities have defined "frameworks" for business modeling. These frameworks attempt to define a rigorous approach to defining business value streams. It is not clear, however, to what extent such frameworks are actually important for business planning. Business model frameworks represent the core aspect of any company; they involve "the totality of how a company selects its customers defines and differentiates its offerings, defines the tasks it will perform itself and those it will outsource, configures its resource, goes to market, creates utility for customers, and captures profits". A business framework involves internal factors (market analysis; products/services promotion; development of trust; social influence and knowledge sharing) and external factors (competitors and technological aspects).

A review on business model frameworks can be found in Krumeich et al. (2012). In the following some frameworks are introduced.

- Business reference model

  Business reference model is a reference model, concentrating on the architectural aspects of the core business of an enterprise, service organization or government agency.

- Component business model

  Technique developed by IBM to model and analyze an enterprise. It is a logical representation or map of business components or "building blocks" and can be depicted on a single page. It can be used to analyze the alignment of enterprise strategy with the organization's capabilities and investments, identify redundant or overlapping business capabilities, etc.

Although Webvan failed in its goal of disintermediating the North American supermarket industry, several supermarket chains (like Safeway Inc.) have launched their own delivery services to target the niche market to which Webvan catered.

- Industrialization of services business model

  Business model used in strategic management and services marketing that treats service provision as an industrial process, subject to industrial optimization procedures

- Business Model Canvas

  Developed by A. Osterwalder, Yves Pigneur, Alan Smith, and 470 practitioners from 45 countries, the business model canvas is one of the most used frameworks for describing the elements of business models.

- OGSM

  The OGSM is developed by Marc van Eck and Ellen van Zanten of Business Openers into the 'Business plan on 1 page'. Translated in several languages all over the world. #1 Management book in The Netherlands in 2015. The foundation of Business plan on 1 page is the OGSM. Objectives, Goals, Strageties and Measures (dashboard and actions).

Related Concepts

The process of business model design is part of business strategy. Business model design and innovation refer to the way a firm (or a network of firms) defines its business logic at the strategic level.

In contrast, firms implement their business model at the operational level, through their business operations. This refers to their process-level activities, capabilities, functions and infrastructure (for example, their business processes and business process modeling), their organisational structures (e.g. organigrams, workflows, human resources) and systems (e.g. information technology architecture, production lines).

Consequently, an operationally viable and feasible business model requires lateral alignment with the underlining business operations.

The brand is a consequence of the business model and has a symbiotic relationship with it, because the business model determines the brand promise, and the brand equity becomes a feature of the model. Managing this is a task of integrated marketing.

The standard terminology and examples of business models do not apply to most non-profit organizations, since their sources of income are generally not the same as the beneficiaries. The term 'funding model' is generally used instead.

The model is defined by the organization's vision, mission, and values, as well as sets of boundaries for the organization—what products or services it will deliver, what customers or markets it will target, and what supply and delivery channels it will use. While the business model includes high-level strategies and tactical direction for how the organization will implement the model, it also includes the annual goals that set the specific steps the organization intends to undertake in the next year and the measures for their expected accomplishment. Each of these is likely to be part of internal documentation that is available to the internal auditor.

Bricks and Clicks

A Sports Direct storefront advertising the web arm of the business. Sports Direct started trading in 1982 with a single brick-and-mortar store but has recently grown rapidly aided by a bricks and clicks business model.

Bricks and clicks (aka clicks and bricks, click and mortar, bricks, clicks and flips, Womble Store Method (WSM) or WAMBAM) is a jargon term for a business model by which a company integrates both offline (*bricks*) and online (*clicks*) presences, sometimes with the third extra *flips* (physical catalogs). Additionally, many will also offer telephone ordering and mobile phone apps, or at least provide telephone sales support. The advent of mobile web has made businesses operating bricks and clicks businesses especially popular, because it means customers can do tasks like shopping when they have spare time and do not have to be at a computer. Many of these users prefer to use mobile shopping sites.

A popular example of the *bricks and clicks* model is when a chain of stores allows the customer to order products either online or physically in one of their stores, also allowing them to either pick-up their order directly at a local branch of the store or get it delivered to their home. There are many alternative combinations of this model. The success of the model in many sectors has lessened the credibility of some analysts who argued that the Internet would render traditional retailers obsolete through disintermediation.

History

A Tesco delivery van in Poland advertising online ordering and delivery from a brick-and-mortar store. Tesco started their online presence in 1996.

The first ever purchase from a company arguably operating a bricks and clicks business model was a Pizza Hut pizza, ordered over the internet from a physical store. The online pizza delivery industry is something of a pioneer of the model and has gained a great deal of popularity since, with delivery company Dominos Pizza now reporting that over 69.7% of orders are placed online before being sent to a physical store, gaining the firm £204.7m (approx. $329m) in 2013 in the United Kingdom alone. The great surge in adoption of the bricks and clicks model came around 2000, with large retailers such as Wal Mart starting websites that allow users to browse the same goods they would find in store from the comfort of their homes.

Advantages

Advantages for Firms

The bricks and clicks model has typically been used by traditional retailers who have extensive logistics and supply chains, but are well known and often respected for their traditional physical presence. Part of the reason for its success is that it is far easier for a traditional retailer to establish an online presence than it is for a start-up company to employ a successful purely online one, or for an online only retailer to establish a traditional presence, including a strong and well recognised brand, without having a large marketing budget. It can also be said that adoption of a bricks and clicks model where a customer can return items to a brick and mortar store can reduce wasted costs to a business such as shipping for undelivered and returned items that would traditionally be incurred.

Advantages for Consumers

A bricks and clicks business model can benefit various members of a customer base. For example, supermarkets often have different customer types requiring alternative shopping options; one group may wish to see the goods directly before purchase and like the convenience of quickly shopping on-the-fly, while another group may require a different convenience of shopping online and getting the order delivered when it suits them, having a bricks and clicks model means both customer groups are satisfied. Other previously online-only retailers have stated that they have found benefit in adding a brick-and-mortar presence to their online-only business, as customers can physically see and test products before purchase as well as get advice and support on any purchases they have made. Additionally, consumers are likely to feel safer and have more confidence using a bricks-and-clicks business if they already know the brand from a brick-and-mortar store.

Disadvantages

Disadvantages for Firms

A major factor in the success or failure of this business model is in the control of costs, as usually maintaining a physical presence —paying for many physical store premises and their staffing— requires larger capital expenditure which online only businesses do not usually have. Conversely, a business selling more luxurious, often expensive, or only occasionally purchased products —like cars— may find sales are more common with a physical presence, due to the more considered nature of the purchasing decision, though they may still offer online product information. However, some car manufacturers such as Dacia have introduced online configurators that allow a customer to configure and order complete cars online, only going to a dealership to collect the completed car, which has proven popular with customers.

"On the other hand, an online-only service can remain a best-in-class operation because its executives focus on just the online business." It has been argued that a bricks and clicks business model is more difficult to implement than an online only model. In the future, the bricks and clicks model may be more successful, but in 2010 some online only businesses grew at a staggering 30%, while some bricks and clicks businesses grew at a paltry 3%. The key factor for a bricks and clicks business model to be successful "will, to a large extent, be determined by a company's ability to manage the trade-offs between separation and integration" of their retail and online businesses.

Disadvantages for Consumers

- Some argue that online shopping, which makes price comparison easier for customers, encourages a 'race-to-the-bottom', where retailers only compete on price, with quality and service deteriorating as a result. This is especially prevalent when comparison shopping websites such as mySupermarket allow prices to be compared without even visiting a retailer's website.
- The prices listed online may not match the prices listed offline. The reasons for this include mis-management, and economics (overhead cost of an online purchase and an offline purchase is different). This may result in confusion and deviations of expectations for the buyers.
- Buyers may end up buying more items than they need, because online businesses are able to show them more items, more promotions, and more advertisements.

Legislation

An advantage to the consumer and a potential disadvantage to businesses is that by adopting a bricks and clicks business model and allowing customers to purchase goods or services remotely, it is legislated in many jurisdictions that consumers are granted more rights to protect them. In the UK, for example, any goods purchased from a bricks and clicks business over a 'click and collect' service would allow the buyer protection under the Consumer Protection (Distance Selling) Regulations 2000, namely the right to return a product or cancel a service within 14 days of purchase for a full refund. Similar rights are afforded to EU Residents, who gain protection under European Directive 97/7/EC. In the USA, the Federal Trade Commission legislate specifically over how a distance sale should be conducted and the rights that a consumer has, namely a '3 day' rule allowing items ordered over the web to be returned within three days.

An example of a retailer falling foul of this legislation is British clothing retailer Next, who were found to be breaking the laws by only allowing a customer to return goods that they had ordered if they paid return postage costs.

Notable Examples

British retailer John Lewis has found success in adopting a bricks and clicks business model, with the online ordering service outperforming brick and mortar sales for several years running. Online auction website eBay have also launched a scheme in cooperation with catalogue shop Argos that allows goods sold by third parties to be collected in a brick-and-mortar location, which allows the customer to collect goods at their convenience rather than wait at home for a delivery company.

In 2013, music retailer HMV went in to administration despite having operated both brick-and-mortar stores and an online presence. This was put down by many to the high overheads of operating the brick-and-mortar side of the business making HMV unable to compete with pure-clicks retailers such as Amazon.com.

Business Model Pattern

Business model patterns are reusable business model architectural components, which can be used in generating a new business model. In the process of new business model generation, the business model innovator can use one or more of these patterns to creating a new business model. Each of these patterns has similarities in characteristics, business model building blocks arrangements and behaviors. Alexander Osterwalder call these similarities the "business model pattern".

Types

- Unbundling business model
- Long tail model
- Multi-sided platform model
- Free as a business model
- Open business model

Loyalty Business Model

The loyalty business model is a business model used in strategic management in which company resources are employed so as to increase the loyalty of customers and other stakeholders in the expectation that corporate objectives will be met or surpassed. A typical example of this type of model is: quality of product or service leads to customer satisfaction, which leads to customer loyalty, which leads to profitability.

The Service Quality Model

A model by Kaj Storbacka, Tore Strandvik, and Christian Grönroos (1994), the service quality model, is more detailed than the basic loyalty business model but arrives at the same conclusion. In it, customer satisfaction is first based on a recent experience of the product or service. This assessment depends on prior expectations of overall quality compared to the actual performance received. If the recent experience exceeds prior expectations, customer satisfaction is likely to be high. Customer satisfaction can also be high even with mediocre performance quality if the customer's expectations are low, or if the performance provides value (that is, it is priced low to reflect the mediocre quality). Likewise, a customer can be dissatisfied with the service encounter and still perceive the overall quality to be good. This occurs when a quality service is priced very high and the transaction provides little value.

This model then looks at the strength of the business relationship; it proposes that this strength is determined by the level of satisfaction with recent experience, overall perceptions of quality, customer commitment to the relationship, and bonds between the parties. Customers are said to have a "zone of tolerance" corresponding to a range of service quality between "barely adequate" and "exceptional." A single disappointing experience may not significantly reduce the strength of the business relationship if the customer's overall perception of quality remains high, if switching costs are high, if there are few satisfactory alternatives, if they are committed to the relationship, and if there are bonds keeping them in the relationship. The existence of these bonds acts as an exit barrier. There are several types of bonds, including: legal bonds (contracts), technological bonds (shared technology), economic bonds (dependence), knowledge bonds, social bonds, cultural or ethnic bonds, ideological bonds, psychological bonds, geographical bonds, time bonds, and planning bonds.

This model then examines the link between relationship strength and customer loyalty. Customer loyalty is determined by three factors: relationship strength, perceived alternatives and critical episodes. The relationship can terminate if:

1. the customer moves away from the company's service area,
2. the customer no longer has a need for the company's products or services,
3. more suitable alternative providers become available,
4. the relationship strength has weakened,
5. the company handles a critical episode poorly,
6. unexplainable change of price of the service provided.

The final link in the model is the effect of customer loyalty on profitability. The fundamental assumption of all the loyalty models is that keeping existing customers is less expensive than acquiring new ones. It is claimed by Reichheld and Sasser (1990) that a 5% improvement in customer retention can cause an increase in profitability between 25% and 85% (in terms of net present value) depending upon the industry. However, Carrol and Reichheld (1992) dispute these calculations, claiming that they result from faulty cross-sectional analysis.

According to Buchanan and Gilles (1990), the increased profitability associated with customer retention efforts occurs because:

- The cost of acquisition occurs only at the beginning of a relationship: the longer the relationship, the lower the amortized cost.
- Account maintenance costs decline as a percentage of total costs (or as a percentage of revenue).
- Long term customers tend to be less inclined to switch and also tend to be less price sensitive. This can result in stable unit sales volume and increases in sales volume.
- Long term customers may initiate free word of mouth promotions and referrals.
- Long term customers are more likely to purchase ancillary products and high-margin supplemental products.
- Long term customers tend to be satisfied with their relationship with the company and are less likely to switch to competitors, making market entry or competitors' market share gains difficult.
- Regular customers tend to be less expensive to service because they are familiar with the processes involved, require less "education," and are consistent in their order placement.
- Increased customer retention and loyalty makes the employees' jobs easier and more satisfying. In turn, happy employees feed back into higher customer satisfaction in a virtuous circle.

For this final link to hold, the relationship must be profitable. Striving to maintain the loyalty of unprofitable customers is not a viable business model. That is why it is important for marketers to assess the profitability of each of its clients (or types of clients), and terminate those relationships that are not profitable. In order to do this, each customer's "relationship costs" are compared to their "relationship revenue." A useful calculation for this is the patronage concentration ratio. This calculation is hindered by the difficulty in allocating costs to individual relationships and the ambiguity regarding relationship cost drivers.

Expanded Models

Schlesinger and Heskett (1991) added employee loyalty to the basic customer loyalty model. They developed the concepts of "cycle of success" and "cycle of failure". In the cycle of success, an investment in your employees' ability to provide superior service to customers can be seen as a virtuous circle. Effort spent in selecting and training employees and creating a corporate culture in which they are empowered can lead to increased employee satisfaction and employee competence. This will likely result in superior service delivery and customer satisfaction. This in turn will create customer loyalty, improved sales levels, and higher profit margins. Some of these profits can be reinvested in employee development thereby initiating another iteration of a virtuous cycle.

Virtuous Circle

Fred Reichheld (1996) expanded the loyalty business model beyond customers and employees. He looked at the benefits of obtaining the loyalty of suppliers, employees, bankers, customers, distributors, shareholders, and the board of directors.

Duff and Einig (2015) expanded the model to debt issuers and credit ratings agencies to investigate what role commitment plays in issuer-CRA relations.

Satisfaction-profit-chain (SPC) Model

The satisfaction-profit chain is a model that theoretically develops linkages and then enables researchers to test them statistically for a firm using customer data (both from surveys and other sources). The satisfaction-profit-chain was tested in the context of banking industry showing that product and services improvements indeed were associated with customer perceptions, which led to beneficial customer behaviors such as repurchase, and desirable financial outcomes such as increased sales and profitability The satisfaction-profit-chain, as a methodology for managing customer loyalty and firm profitability, is also applicable in business-to-business markets, irrespective of whether the B2B firm sells goods and/or services.

The satisfaction-profit-chain refers to a chain of effects whereby increased performance on key attributes leads to improvements in overall satisfaction, which in turn affects loyalty intentions and behaviors. The increased customer loyalty is shown to affect short- and long-term financial outcomes including sales, profitability, and stock price. More recently, some studies show that especially in the context of services such as retailing and financial services, employee satisfaction can play a critical role in enhancing customer loyalty. This happens because both customer satisfaction and employee satisfaction can mutually reinforce each other, and promote stronger customer loyalty. More specifically, for a given level of overall satisfaction, customer loyalty is disproportionately stronger when customers perceive that employees are also satisfied.

The SPC model has become the basis of a large body of empirical research showing the strong impact of customer satisfaction on customer loyalty. Research has clearly shown that one of the best ways to increase customer loyalty—measured as repurchase intentions and/or repurchase behavior—is by increasing customer satisfaction (more satisfied customers are more loyal, in general). Though the relationship is positive, research shows there are many differences:

1) The effect of customer satisfaction on customer loyalty can vary based on customer demographics and segments, such that it is stronger for some demographic groups and segments than others.

2) The effect of customer satisfaction and customer loyalty, and subsequent financial outcomes for firms, can vary based on industry. Specifically, factors such as—goods versus services industry, degree of competition or concentration in the industry, the utilitarian or hedonic nature of products, and customers' switching costs can affect the nature (non-linearity) and strength of the link between customer satisfaction and customer loyalty.

3) The measurement of loyalty—especially for customers is multi-faceted. Customer loyalty includes a variety of outcomes—intentions and behaviors associated with repurchase including word-of-mouth, complaint behaviors, share-of-wallet or the relative proportion of purchasing from a single firm relative to customer's total purchasing, and likelihood to recommend.

4) Customer loyalty is influenced, not only by customer satisfaction but also employee satisfaction. Customer loyalty is a function of customer satisfaction. In many firms, especially service-oriented industries such as retailing, health-care, financial services, education, and hospitality the level of satisfaction experienced by front-line employees is a critical component. The level of employee satisfaction influences customer satisfaction as shown in a large-scale study of managers, front-line employees, and customers of a DIY retailer in Europe: results showed that managers affected overall job-satisfaction of front-line employees, which in turn affected the satisfaction of customers they interacted with. Most surprisingly, the level of customer loyalty was much higher among those customers who were themselves more satisfied, but also interacted with more satisfied employees. Highly satisfied customers who dealt with relatively less satisfied employees were relatively less loyal.

Commitment-loyalty Model

The customer commitment approach to loyalty is based on the idea that customers with higher commitment toward the brand are also more likely to be loyal toward the brand. Earlier models of customer commitment conceptualized it as a unidimensional construct (e.g., Garbarino and Johnson 1999; Moorman et al. 1992). More recently, scholars have developed a five dimensional scale to measure customer commitment and relate it to customer loyalty. The five commitment dimensions include:

- Affective commitment

- Normative commitment
- Economic commitment
- Forced commitment
- Habitual commitment

Data Collection

Typically, loyalty data is being collected by multi-item measurement scales administered in questionnaires by software providers such as Confirmit, Medallia, and Satmetrix. However, other approaches sometimes seem more viable if managers want to know the extent of loyalty for an entire data warehouse. This approach is described in Buckinx, Verstraeten & Van den Poel (2006).

All historical trends for different segmentations and their standard of living may also be very helpful in developing customer retention strategy. Lifestyle is also a very powerful tool, can be used for better customer retention and to know his/her needs in better way.

Industrialization of Services Business Model

The industrialization of services business model is a business model used in strategic management and services marketing that treats service provision as an industrial process, subject to industrial optimization procedures. It originated in the early 1970s, at a time when various quality control techniques were being successfully implemented on production assembly lines.

Theodore Levitt (1972) argued that the reason the service sector suffered from inefficiency and wide variations in quality were that it was based on the craft model. Each service encounter was treated as an isolated event. He felt that service encounters could be systematized through planning, optimal processes, consistency, and capital intensive investments. This model was the foundation of the success of McDonalds and many other mass service providers in the 1970s, 80s, and 90s.

Unfortunately, the application of assembly line techniques to service provision had several undesirable consequences. Employees found working under these conditions disempowering, resulting in low morale, high staff turnover, and reduced service quality. One of the most difficult aspects of this model for employees to deal with was the "smile incentives". Employees were instructed to put a smile on their face during the service encounter. This manufacturing and commercialization of apparent happiness has been criticised by many commentators, particularly Mundie (1987). Also many customers prefer the "personal touch".

By the early 1990s most service providers turned their attention back to the human element and personalized their services. Employees were empowered to customize the service encounter to the individual characteristics of customers.

Subsequently, scholars developed the service-profit-chain concept to understand how employees and customers interact to create value.

Dependent Growth Business Model

The Dependent Growth Business Model is a business model employed by a social network service where the rationale of how an organization creates, delivers, and captures value is decided by a particular set of predetermined and commuted factors or circumstances.

In addition, as opposed to an autonomous business model, in a growth dependent business model members may serve as both the suppliers and the consumers of content as well filling each role singly. The growth business model also details specific qualities in a supportive work environment that promotes and employs sustainability and environmentally friendly operational standards.

Pyramid Scheme

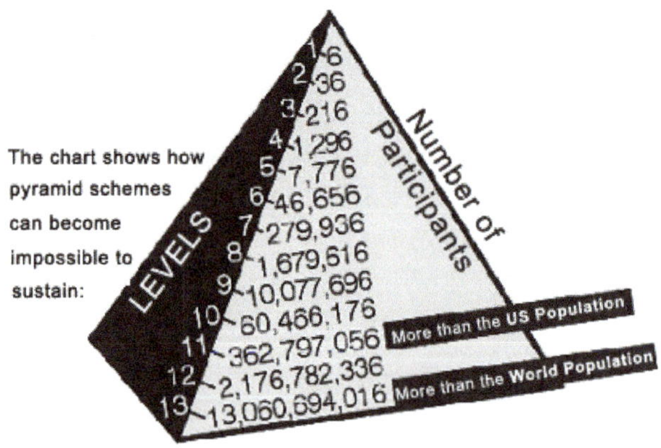

The unsustainable exponential progression of a classic pyramid scheme (here with a branching factor of 6)

A pyramid scheme is a business model that recruits members via a promise of payments or services for enrolling others into the scheme, rather than supplying investments or sale of products or services. As recruiting multiplies, recruiting becomes quickly impossible, and most members are unable to profit; as such, pyramid schemes are unsustainable and often illegal.

Pyramid schemes have existed for at least a century in different guises. Some multilevel marketing plans have been classified as pyramid schemes.

Concept and Basic Models

In a pyramid scheme, an organization compels individuals to join and make a payment. In exchange, the organization promises its new members a share of the money taken from every additional member that they recruit. The directors of the organization (those at the top of the pyramid) also receive a share of these payments. For the directors, the scheme is potentially lucrative—whether or not they do any work, the organization's membership has a strong incentive to continue recruiting and funneling money to the top of the pyramid.

Such organizations seldom involve sales of products or services with value. Without creating any goods or services, the only ways for a pyramid scheme to generate revenue are to recruit more members or solicit more money from current members. Eventually, recruiting is no longer possible and most members are unable to profit from the scheme.

The "Eightball" Model

Many pyramids are more sophisticated than the simple model. These recognize that recruiting a large number of others into a scheme can be difficult so a seemingly simpler model is used. In this model each person must recruit two others, but the ease of achieving this is offset because the depth required to recoup any money also increases. The scheme requires a person to recruit two others, who must each recruit two others, and so on.

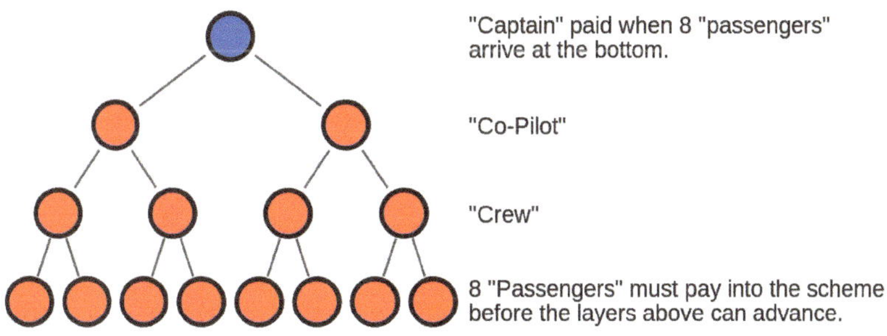

The "eight-ball" model contains a total of fifteen members. Note that in an arithmetic progression 1 + 2 + 3 + 4 + 5 = 15. The pyramid scheme in the picture in contrast is a geometric progression 1 + 2 + 4 + 8 = 15.

Prior instances of this scheme have been called the "Airplane Game" and the four tiers labelled as "captain", "co-pilot", "crew", and "passenger" to denote a person's level. Another instance was called the "Original Dinner Party" which labeled the tiers as "dessert", "main course", "side salad", and "appetizer". A person on the "dessert" course is the one at the top of the tree. Another variant, "Treasure Traders", variously used gemology terms such as "polishers", "stone cutters", etc.

Such schemes may try to downplay their pyramid nature by referring to themselves as "gifting circles" with money being "gifted". Popular schemes such as "Women Empowering Women" do exactly this.

Whichever euphemism is used, there are 15 total people in four tiers (1 + 2 + 4 + 8) in the scheme—with the Airplane Game as the example, the person at the top of this tree is the "captain", the two below are "co-pilots", the four below are "crew," and the bottom eight joiners are the "passengers".

The eight passengers must each pay (or "gift") a sum (e.g., $5,000) to join the scheme. This sum (e.g., $40,000) goes to the captain who leaves, with everyone remaining moving up one tier. There are now two new captains so the group splits in two with each group requiring eight new passengers. A person who joins the scheme as a passenger will not see a return until they advance through the crew and co-pilot tiers and exit the scheme as a captain. Therefore, the participants in the bottom three tiers of the pyramid lose their money if the scheme collapses.

If a person is using this model as a scam, the confidence trickster would take the majority of the money. They would do this by filling in the first three tiers (with one, two, and four people) with phony names, ensuring they get the first seven payouts, at eight times the buy-in sum, without paying a single penny themselves. So if the buy-in were $5,000, they would receive $40,000, paid for by the first eight investors. They would continue to buy in underneath the real investors, and promote and prolong the scheme for as long as possible to allow them to skim even more from it before it collapses.

Although the "captain" is the person at the top of the tree, having received the payment from the eight paying passengers, once they leave the scheme they are able to re-enter the pyramid as a "passenger" and hopefully recruit enough to reach captain again, thereby earning a second payout.

Matrix Schemes

Matrix schemes use the same fraudulent non-sustainable system as a pyramid; here, the participants pay to join a waiting list for a desirable product, which only a fraction of them can ever receive. Since matrix schemes follow the same laws of geometric progression as pyramids, they are subsequently as doomed to collapse. Such schemes operate as a queue, where the person at head of the queue receives an item such as a television, games console, digital camcorder, etc. when a certain number of new people join the end of the queue. For example, ten joiners may be required for the person at the front to receive their item and leave the queue. Each joiner is required to buy an expensive but potentially worthless item, such as an e-book, for their position in the queue. The scheme organizer profits because the income from joiners far exceeds the cost of sending out the item to the person at the front. Organizers can further profit by starting a scheme with a queue with *shill* names that must be cleared out before genuine people get to the front. The scheme collapses when no more people are willing to join the queue. Schemes may not reveal, or may attempt to exaggerate, a prospective joiner's queue position, a condition that essentially means the scheme is a lottery. Some countries have ruled that matrix schemes are illegal on that basis.

Relation to Ponzi Schemes

While often confused for each other, Pyramid schemes and Ponzi schemes are different from each other. They are related in the sense that both pyramid and Ponzi schemes are forms of financial fraud. However, Pyramid schemes are based on network marketing, where each part of the pyramid takes a piece of the pie / benefits, forwarding the money to the top of the pyramid. They fail simply because there aren't sufficient people, hence by the time one comes to the tenth level, the pyramid needs 10 billion people to sustain it. Ponzi schemes, on the other hand are based on the principle of "Robbing Peter to pay Paul" - early investors are paid their returns through the proceeds of investments by later investors. In other words, one central person (or entity) in the middle taking money from one person, keeping part of it and giving the rest to others who had invested in the scheme earlier. Thus, schemes such as the Anubhav teak plantation scheme (Teak plantation scam of 1998) in India can be called Ponzi schemes. Some Ponzi schemes can depend on multi level marketing for popularizing them, thus forming a combination of the two.

Connection to Multi-level Marketing

Some multi-level marketing (MLM) companies operate as pyramid schemes and consumers often confuse legitimate multi-level marketing with pyramid schemes.

According to the U.S. Federal Trade Commission legitimate MLM, unlike pyramid schemes:

> "have a real product to sell. More importantly, MLM's [*sic*] actually sell their product to members of the general public, without requiring these consumers to pay anything extra or to join the MLM system. MLM's may pay commissions to a long string of distributors, but these commission are paid for real retail sales, not for new recruits."

Pyramid schemes however "may purport to sell a product, but they often simply use the product to hide their pyramid structure". While some people call MLMs in general "pyramid selling" others use the term to denote an illegal pyramid scheme masquerading as an MLM.

The Federal Trade Commission warns, "It's best not to get involved in plans where the money you make is based primarily on the number of distributors you recruit and your sales to them, rather than on your sales to people outside the plan who intend to use the products." It states that research is your best tool and gives eight steps to follow:

1. Find—and study—the company's track record.
2. Learn about the product.
3. Ask questions.
4. Understand any restrictions.
5. Talk to other distributors. Beware of shills.
6. Consider using a friend or adviser as a neutral sounding board, or for a gut check.
7. Take your time.
8. Think about whether this plan suits your talents and goals.

Some commentators contend that MLMs in general are nothing more than legalized pyramid schemes.

Legality

Pyramid schemes are illegal in many countries or regions including Albania, Australia, Austria, Belgium, Brazil, Canada, China, Colombia, Denmark, the Dominican Republic, Estonia, Finland, France, Germany, Hong Kong, Hungary, Iceland, Iran, the Republic of Ireland, Italy, Japan, Malaysia, Maldives, Mexico, Nepal, the Netherlands, New Zealand, Norway, the Philippines, Poland, Portugal, Romania, Russian Federation, Serbia, South Africa, Spain, Sri Lanka, Sweden, Switzerland, Taiwan, Thailand, Turkey, Ukraine, the United Kingdom, and the United States.

Franchise fraud is defined by the United States Federal Bureau of Investigation as a pyramid scheme. The FBI website states:

Pyramid schemes—also referred to as franchise fraud or chain referral schemes—are marketing and investment frauds in which an individual is offered a distributorship or franchise to market a particular product. The real profit is earned, not by the sale of the product, but by the sale of new distributorships. Emphasis on selling franchises rather than the product eventually leads to a point where the supply of potential investors is exhausted and the pyramid collapses.

Notable Recent Cases

The 1997 rebellion in Albania was partially motivated by the collapse of Ponzi schemes; however, they were widely referred to as pyramid schemes due to their prevalence in Albanian society.

In 2003, the United States Federal Trade Commission (FTC) disclosed what it called an Internet-based "pyramid scam." Its complaint states that customers would pay a registration fee to join a program that called itself an "internet mall" and purchase a package of goods and services such as internet mail, and that the company offered "significant commissions" to consumers who purchased and resold the package. The FTC alleged that the company's program was instead and in reality a pyramid scheme that did not disclose that most consumers' money would be kept, and that it gave affiliates material that allowed them to scam others.

WinCapita was a scheme run by Finnish criminals that involved about €100 million. The scheme started in 2005.

In early 2006, Ireland was hit by a wave of schemes with major activity in Cork and Galway. Participants were asked to contribute €20,000 each to a "Liberty" scheme which followed the classic eight-ball model. Payments were made in Munich, Germany to skirt Irish tax laws concerning gifts. Spin-off schemes called "Speedball" and "People in Profit" prompted a number of violent incidents and calls were made by politicians to tighten existing legislation. Ireland has launched a website to better educate consumers to pyramid schemes and other scams.

On 12 November 2008, riots broke out in the municipalities of Pasto, Tumaco, Popayan and Santander de Quilichao, Colombia after the collapse of several pyramid schemes. Thousands of victims had invested their money in pyramids that promised them extraordinary interest rates. The lack of regulation laws allowed those pyramids to grow excessively during several years. Finally, after the riots, the Colombian government was forced to declare the country in a state of economic emergency to seize and stop those schemes. Several of the pyramid's managers were arrested, and are being prosecuted for the crime of "illegal massive money reception."

The *Kyiv Post* reported on 26 November 2008 that American citizen Robert Fletcher (Robert T. Fletcher III; aka "Rob") was arrested by the SBU (Ukraine State Police) after being accused by Ukrainian investors of running a Ponzi scheme and associated pyramid scam netting US$20 million. (The *Kiev Post* also reports that some estimates are as high as US$150M.)

In the United Kingdom in 2008 and 2009, a £21 million pyramid scheme named 'Give and Take' involved at least 10,000 victims in the south-west of England and South Wales. Leaders of the scheme were prosecuted and served time in jail before being ordered to pay £500,000 in compensation and costs in 2015. The cost of bringing the prosecution was in excess of £1.4 million.

Throughout 2010 and 2011 a number of authorities around the world including the Australian Competition and Consumer Commission, the Bank of Namibia and the Central Bank of Lesotho have declared TVI Express to be a pyramid scheme. TVI Express, operated by Tarun Trikha from India has apparently recruited hundreds of thousands of "investors", very few of whom, it is reported, have recouped any of their investment. In 2013, Tarun Trikha was arrested at the IGI Airport in New Delhi.

BurnLounge, Inc. was a multi-level marketing online music store founded in 2004 and based in New York City. By 2006 the company reported 30,000 members using the site to sell music through its network. In 2007 the company was sued by the Federal Trade Commission for being an illegal pyramid scheme. The company lost the suit in 2012, and lost appeal in June 2014. In June 2015, the FTC began returning $1.9 million to people who had lost money in the scheme.

Business Models for Open-source Software

Open-source software is widely used both as independent applications and as components in non-open-source applications. Many independent software vendors (ISVs), value-added resellers (VARs), and hardware vendors (OEMs or ODMs) use open-source frameworks, modules, and libraries inside their proprietary, for-profit products and services. From a customer's perspective, the ability to use open technology under standard commercial terms and support is valuable. They are willing to pay for the legal protection (e.g., indemnification from copyright or patent infringement), "commercial-grade QA", and professional support/training/consulting that are typical of commercial software, while also receiving the benefits of fine-grained control and lack of lock-in that comes with open-source.

Funding

Much unlike proprietary off-the-shelf software that come with restrictive licenses, open-source software is distributed freely, through the web and in physical media. Because creators cannot require each user to pay a license fee to fund development this way, a number of alternative development funding models have emerged.

An example of those funding models is when bespoke software is developed as a consulting project for one or more customers who request it. These customers pay developers to have this software developed according to their own needs and they could also closely direct the developers' work. If both parties agree, the resulting software could then be publicly released with an open-source license in order to allow subsequent adoption by other parties. That agreement could reduce the costs paid by the clients while the original developers (or independent consultants) can then charge for training, installation, technical support, or further customization if and when more interested customers would choose to use it after the initial release.

There also exist stipends to support the development of open source software, such as Google's Summer of Code.

Another approach to funding is to provide the software freely, but sell licenses to proprietary add-ons such as data libraries. For instance, an open-source CAD program may require parts libraries which are sold on a subscription or flat-fee basis. Open-source software can also promote the sale of specialized hardware that it interoperates with, some example cases being the Asterisk telephony software developed by PC-telephony hardware manufacturer Digium and the Robot Operating System (ROS) robotics platform by Willow Garage and Stanford AI Labs. Many open source software projects have begun as research projects within universities, as personal projects of students or professors, or as tools to aid scientific research. The influence of universities and research institutions on open-source shows in the number of projects named after their host institutions, such as BSD Unix, CMU Common Lisp, or the NCSA HTTPd which evolved into Apache.

Companies may employ developers to work on open-source projects that are useful to the company's infrastructure: in this case, it is developed not as a product to be sold but as a sort of shared public utility. A local bug-fix or solution to a software problem, written by a developer either at a company's request or to make his/her own job easier, can be released as an open-source contribution without costing the company anything. A larger project such as the Linux kernel may have contributors from dozens of companies which use and depend upon it, as well as hobbyist and research developers.

A new funding approach for open-source projects is crowdfunding, organized over web platforms like Kickstarter, Indiegogo, or Bountysource.

Challenges

Open-source software can be sold and used in general commercially. Also, commercial open-source applications are a part of the software industry for some time. While commercialization or funding of open-source software projects is possible, it is considered challenging.

Since several open-source licenses stipulate that authors of derivative works must distribute them under an open-source (copyleft) license, ISVs and VARs have to develop new legal and technical mechanisms to foster their commercial goals, as many traditional mechanisms are not directly applicable anymore.

Traditional business wisdom suggests that a company's methods, assets, and intellectual properties should remain concealed from market competitors (trade secret) as long as possible to maximize the profitable commercialization time of a new product. Open-source software development minimizes the effectiveness of this tactic; development of the product is usually performed in view of the public, allowing competing projects or clones to incorporate new features or improvements as soon as the public code repository is updated, as permitted by most open-source licenses. Also in the computer hardware domain, a hardware producer who provides free and open software drivers reveals the knowledge about hardware implementation details to competitors, who might use this knowledge to catch up.

Therefore, there is considerable debate about whether vendors can make a sustainable business from an open-source strategy. In terms of a traditional software company, this is probably the wrong question to ask. Looking at the landscape of open source applications, many of the larger ones are sponsored (and largely written) by system companies such as IBM who may not have an objective of software license revenues. Other software companies, such as Oracle and Google, have sponsored or delivered significant open-source code bases. These firms' motivation tends to be more strategic, in the sense that they are trying to change the rules of a marketplace and reduce the influence of vendors such as Microsoft. Smaller vendors doing open-source work may be less concerned with immediate revenue growth than developing a large and loyal community, which may be the basis of a corporate valuation at merger time.

Approaches

A variety of open-source compatible business approaches have gained prominence in recent years; notable examples include dual licensing, software as a service, not charging for the software but for services, freemium, donation-based funding, and crowdfunding.

There are several different types of business models for making profit using open-source software (OSS) or funding the creation. Below are existing and legal commercial business approaches in context of open-source software and open-source licenses. The acceptance of these approaches varies; some of these approaches are recommended (like selling services), others are accepted, while still others are considered controversial or even unethical by the open-source community. The underlying objective of these business models is to harness the size and international scope of the open-source community (typically more than an order of magnitude larger than what would be achieved with closed-source models) for a sustainable commercial venture. The vast majority of commercial open-source companies experience a conversion ratio (as measured by the percentage of downloaders who buy something) well below 1%, so low-cost and highly-scalable marketing and sales functions are key to these firms' profitability.

Dual-licensing

Dual licensing offers the software under an open-source license but also under separate proprietary license terms. The proprietary version can be sold to finance the continued development of the free open-source version. Customers can be attracted to a no-cost and open-source edition, then be part of an up-sell to a commercial enterprise edition. Further, customers will learn of open-source software in a company's portfolio and offerings but generate business in other proprietary products and solutions, including commercial technical support contracts and services. A popular example is Oracle's MySQL database which is dual-licensed under a commercial proprietary license as also under the GPLv2. Another example is the Sleepycat License. Flask developer Armin Ronacher stated that the AGPLv3 was a "terrible success" as "vehicle for dual commercial licensing" and noted that MongoDB, RethinkDB, OpenERP, SugarCRM as well as WURFL utilizing the license for this purpose.

Selling Professional Services

The financial return of costs on open-source software can also come from selling services, such as training, technical support, or consulting, rather than the software itself.

Another possibility is offering open-source software in source code form only, while providing executable binaries to paying customers only, offering the commercial service of compiling and packaging of the software. Also, providing goods like physical installation media (e.g., DVDs) can be a commercial service.

Open-source companies using this business model successfully are for instance RedHat and IBM; a more specialized example is that of Revolution Analytics.

Selling of Branded Merchandise

Some open-source organizations such as the Mozilla Foundation and the Wikimedia Foundation sell branded merchandise articles like t-shirts and coffee mugs. This can be also seen as an additional service provided to the user community.

Selling of Certificates and Trademark Use

Another financing approach is innovated by Moodle, an open source learning management system and community platform. The business model revolves around a network of commercial partners who are certificated and therefore authorised to use the Moodle name and logo, and in turn provide a proportion of revenue to the Moodle Trust, which funds core development.

Selling Software as a Service

Selling subscriptions for online accounts and server access to customers is a way of making profit based on open-source software. Also, combining desktop software with a service, called software plus services. Providing cloud computing services or software as a service (SaaS) without the release of the open-source software itself, neither in binary nor in source form, conforms with most open-source licenses (with exception of the AGPL).

Because of its lack of software freedoms, Richard Stallman calls SaaS "inherently bad" while acknowledging its legality. The FSF called the server-side use-case without release of the source-code the "ASP loophole in the GPLv2" and encourage therefore the use of the Affero General Public License which plugged this hole in 2002. In 2007 the FSF contemplated including the special provision of AGPLv1 into GPLv3 but ultimately decided to keep the licenses separate.

Partnership with Funding Organizations

Other financial situations include partnerships with other companies. Governments, universities, companies, and non-governmental organizations may develop internally or hire a contractor for custom in-house modifications, then release that code under an open-source license. Some organizations support the development of open-source software by grants or stipends, like Google's Summer of Code initiative founded in 2005.

Voluntary Donations

Also, there were experiments by Independent developers to fund development of open-source software donation-driven directly by the users, e.g., with the Illumination Software Creator in 2012. SourceForge, for example, allows users to donate money to the projects it hosts that opt to accept donations. Internet micro-payments systems like PayPal, flattr, and Bitcoin help this approach.

Larger donation campaigns also exist. In 2004 the Mozilla Foundation carried out a fundraising campaign to support the launch of the Firefox 1.0 web browser. It placed a two-page ad in the December 16 edition of the *New York Times* listing the names of the thousands who had donated.

Bounty Driven Development

The users of a particular software artifact may come together and pool money into an open-source bounty for the implementation of a desired feature or functionality. Offering bounties as funding has existed for some time. For instance, Bountysource is a web platform which has offered this funding model for open source software since 2003.

Another bounty source is companies or foundations that set up bounty programs for implemented features or bugfixes in open-source software relevant to them. For instance, Mozilla has been paying and funding freelance open-source programmers for security bug hunting and fixing since 2004.

Pre-order/Crowdfunding/Reverse-bounty Model

A newer funding opportunity for open-source software projects is crowdfunding, which shares similarities with the pre-order or Praenumeration business model, as well as the reverse bounty model. It is typically organized over web platforms like Kickstarter, Indiegogo, or Bountysource. An example is a successfully funded Indiegogo campaign of Australian programmer Timothy Arceri, who offered for $2,500 to implement in two weeks an OpenGL 4.3 extension for the Mesa library. Arceri delivered the OpenGL extension code, which got merged into Mesa, and continued later his Mesa work with successive crowdfunding campaigns.

Advertising-supported Software

In order to commercialize FOSS, many companies (including Google, Mozilla, and Canonical) have moved towards an economic model of advertising-supported software. For instance, the open-source application AdBlock Plus gets paid by Google for letting whitelisted *Acceptable Ads* bypassing the browser ad remover. As another example is SourceForge, an open-source project service provider, has the revenue model of advertising banner sales on their website. In 2006, SourceForge reported quarterly takings of $6.5 million and $23 million in 2009.

Selling of Optional Proprietary Extensions

Some companies sell proprietary but optional extensions, modules, plugins or add-ons to an open-source software product. This can be a "license conform" approach with many open-source licenses if done technically sufficiently carefully. For instance, mixing proprietary code and open-source licensed code in statically linked libraries or compiling all source code together in a software product might violate open-source licenses, while keeping them separated by interfaces and dynamic-link libraries might often adhere to license conform.

This approach is a variant of the freemium business model. The proprietary software may be intended to let customers get more value out of their data, infrastructure, or platform, e.g., operate their infrastructure/platform more effectively and efficiently, manage it better, or secure it better. Examples include the IBM proprietary Linux software, where IBM contributes to the Linux open-source ecosystem, but it builds and delivers (to IBM's paying customers) database software, middleware, and other software that runs on top of the open-source core. Other examples of proprietary products built on open-source software include Red Hat Enterprise Linux and Cloudera's Apache Hadoop-based software. Some companies appear to re-invest a portion of their financial profits from the sale of proprietary software back into the open source infrastructure.

Some companies, such as Digium, sell proprietary but optional digital electronics hardware controlled by an open-source software product.

Selling of Required Proprietary Parts of a Software Product

A variant of the approach above is the keeping of required data content (for instance a video game's audio, graphic, and other art assets) of a software product proprietary while making the software's source code open-source. While this approach is completely legitimate and compatible with most open-source licenses, customers have to buy the content to have a complete and working software product. Restrictive licenses can then be applied on the content, which prevents the redistribution or re-selling of the complete software product. Examples for open-source developed software are Kot-in-Action Creative Artel video game *Steel Storm*, engine GPLv2 licensed while the artwork is CC-BY-NC-SA 3.0 licensed, and *Frogatto & Friends* with an own developed open-source engine and commercialization via the copyrighted game assets for iPhone, BlackBerry and MacOS.

Other examples are *Arx Fatalis* (by Arkane Studios) and *Catacomb 3-D* (by Flat Rock Software) with source code opened to the public delayed after release, while copyrighted assets and binaries are still sold on gog.com as digital distribution.

Doing so conforms with the FSF and Richard Stallman, who stated that for art or entertainment the software freedoms are not required or important.

The similar product bundling of an open-source software product with a proprietary hardware part is called tivoization and legal with most open-source licenses except GPLv3, which explicitly prohibits this use-case.

Re-licensing under a Proprietary License

If a software product uses only own software and open-source software under a permissive free software licence, a company can re-license the resulting software product under a proprietary license and sell the product without the source code or software freedoms. For instance, Apple Inc. is an avid user of this approach by using source code and software from open-source projects. For example, the BSD Unix operating system kernel (under the BSD license) was used in Apple's Mac PCs that were sold as proprietary products.

Obfuscation of Source Code

An approach to allow commercialization under some open-source licenses while still protecting crucial business secrets, intellectual property and technical know-how is obfuscation of source code. This approach was used in several cases, for instance by Nvidia in their open-source graphic card device drivers. This practise is used to get the open-source-friendly propaganda without bearing the inconveniences. There has been debate in the free-software/open-source community on whether it is illegal to skirt copyleft software licenses by releasing source code in obfuscated form, such as in cases in which the author is less willing to make the source code available. The general consensus was that while unethical, it was not considered a violation.

The Free Software Foundation, on the other hand, is clearly against this practice. The GNU General Public License since version 2 has defined "source code" as "the preferred form of the work for making modifications to it." This is intended to prevent the release of obfuscated source code.

Delayed Open-sourcing

Some companies provide the latest version available only to paying customers. A vendor forks a non-copyleft software project then adds closed-source additions to it and sells the resulting software. After a fixed time period the patches are released back upstream under the same license as the rest of the codebase. This business model is called version lagging or time delaying.

For instance, the MariaDB Corporation created for business compatible "delayed open-sourcing" the source-available *Business source license* (BSL) which automatically relicenses after three years to the FOSS GPL. This approach garantuees for licensees that they have source code access (e.g. for code audits), are not locked into a closed platform, or suffer from planned obsolescence, while for the software developer a time-limited exclusive commercialization is possible.

However, this approach works only with own software or permissive licensed code parts, as there is no copyleft FOSS license available which allows the time delayed opening of the source code after distributing or selling of a software product.

Open Sourcing on End-of-life

An extreme variant of "delayed open-sourcing" is a business practice popularized by id Software and 3D Realms, which released several software products under a free software license after a long proprietary commercialization time period and the return of investment was achieved. The motivation of companies following this practice of releasing the source code when a software reaches the commercial end-of-life, is to prevent that their software becomes unsupported Abandonware or even get lost due to digital obsolescence. This gives the user communities the chance to continue development and support of the software product themselves as an open-source software project. Many examples from the video game domain are in the list of commercial video games with later released source code.

Popular non-game software examples are the Netscape Communicator which was open-sourced in 1998 and Sun Microsystems's office suite, StarOffice, which was released in October 2000 at its commercial end of life. Both releases formed the basis of important open-source projects, namely the Mozilla Firefox and OpenOffice.org/LibreOffice. However, Firefox eventually gained a more-than-self-sustaining revenue model, so Firefox was not an example of a commercial end-of-life release.

FOSS and Economy

According to Yochai Benkler, the Berkman Professor for Entrepreneurial Legal Studies at Harvard Law School, free software is the most visible part of a new economy of commons-based peer production of information, knowledge, and culture. As examples, he cites a variety of FOSS projects, including both free software and open source.

This new economy is already under development. In order to commercialize FOSS, many companies, Google being the most successful, are moving towards an economic model of advertising-supported software. In such a model, the only way to increase revenue is to make the advertising more valuable. Facebook has recently come under fire for using novel user tracking methods to accomplish this.

This new economy is not without alternatives. Apple's App Stores have proven very popular with both users and developers. The Free Software Foundation considers Apple's App Stores to be incompatible with its GPL and complained that Apple was infringing on the GPL with its iTunes terms of use. Rather than change those terms to comply with the GPL, Apple removed the GPL-licensed products from its App Stores. The authors of VLC, one of the GPL-licensed programs at the center of those complaints, recently began the process to switch from the GPL to the LGPL and MPL.

Examples

Much of the Internet runs on open-source software tools and utilities such as Linux, Apache, MySQL, and PHP, known as the LAMP stack for web servers. Using open source appeals to software developers for three main reasons: low or no cost, access to source code they can tailor themselves, and a shared community that ensures a generally robust code base, with quick fixes for new issues.

Despite doing much business in proprietary software, some companies like Oracle Corporation and IBM participated in developing free and open-source software to deter from monopolies and take a portion of market share for themselves. See Commercial open-source applications for the list of current commercial open-source offerings. Netscape's actions were an example of this, and thus Mozilla Firefox has become more popular, getting market share from Internet Explorer.

- Active Agenda is offered for free, but requires all extensions to be shared back with the world community. The project sells a "Non-Reciprocal Private License" to anyone interested in keeping module extensions private.
- Adobe Systems offers Flex for free, while selling the Flash Builder IDE.
- Apple Inc. offers Darwin for free, while selling Mac OS X.
- Asterisk (PBX), digital electronics hardware controlled by open-source software
- Codeweavers sells CrossOver commercially, deriving it from the free Wine project they also back.
- Canonical Ltd. offers Ubuntu for free, while they sell commercial technical support contracts.
- Cloudera's Apache Hadoop-based software.
- Francisco Burzi offers PHP-Nuke for free, but the latest version is offered commercially.

- DaDaBIK, although following a donationware approach, requires a small, minimum donation fee, to be downloaded.
- IBM proprietary Linux software, where IBM delivers database software, middleware and other software.
- Ingres is offered for free, but services and support are offered as a subscription. The Ingres Icebreaker Appliance is also offered as a commercial database appliance.
- id Software releases their legacy game engines under the GPL, while retaining proprietary ownership on their latest incarnation.
- Mozilla Foundation have a partnership with Google and other companies which provides revenue for inclusion of search engines in Mozilla Firefox.
- MySQL is offered for free, but with the enterprise version includes support and additional features.
- SUSE offers openSUSE for free through the openSUSE Project, while selling SUSE Linux Enterprise (SLE).
- OpenSearchServer offers its community edition on SourceForge and an enterprise edition with professional services to enterprises with a paid license
- Oracle - VirtualBox is free and open to anyone, but the VirtualBox extension pack can only be used for free at home, thus requiring payment from business users
- OWASP Foundation is a professional community of open-source developers focused on raising visible for software security.
- Red Hat sells support subscriptions for Red Hat Enterprise Linux (RHEL) which is an enterprise distribution periodically forked from the community-developed Fedora.
- Sourcefire offers Snort for free, while selling Sourcefire 3D.
- Sun Microsystems (acquired by Oracle in 2010) once offered OpenOffice.org for free, while selling StarOffice
- Untangle provides its Lite Package for free, while selling its Standard and Premium Packages by subscription
- Zend Technologies offers Zend Server CE and Zend Framework for free, but sells Zend Server with support and additional features.

Inclusive Business Model

An inclusive business model is a commercially viable model that benefits low-income communities by including them in a company's value chain on the demand side as clients and consumers, and/or on the supply side as producers, entrepreneurs or employees in a sustainable way. The concept was first formalized in an early United Nations report called Creating Value for All: Strategies for Doing Business with the Poor (2008) published by the Growing Inclusive Markets Initiative and guided by an Advisory Board consisting of leaders in the field such as the International Business Leaders Forum, the International Finance Corporation, key bilateral donors (USAID and AFD), the World Business Council for Sustainable Development, University of Michigan and Harvard Business School.

Inclusive business models build bridges between business and the poor for mutual benefit. The benefits for business go beyond immediate profits and higher incomes. For business they include driving innovations, building markets and strengthening supply chains. And for the poor they include access to essential goods and services, higher productivity, sustainable earnings and greater empowerment.

The businesses that create and use these innovative models range from multinational corporations to large domestic companies, co-operatives, small and medium-sized enterprises, or even not-for-profit organizations that use business principles—or social business approaches—to achieve their mission.

Inclusive business is not corporate philanthropy or corporate social responsibility, which have inherent limitations of scope, impact and budget. Rather, it is the search for sustainable business models that "do well by doing good" and are part of the companies' core business activities - the key to business having development impact at scale.

Criteria

Inclusive business models can be developed and implemented by a wide range of entities, from private corporations (large and small), to state-owned companies, co-operatives, or even not-for-profit organizations, as long as the following criteria are met:

- *Human development impact*: an inclusive business model contributes to human development by increasing poor people's incomes, improving their access to basic goods and services such as education, health, housing, water and sanitation, contributing to the Millennium Development Goals and reaching excluded and disadvantaged groups (e.g. women, youth, disabled, ethnic minorities).
- *Commercial viability*: an inclusive business model can receive start-up funding from different sources (including grants) but it must be designed to break-even and become self-sustainable over time (profits can be re-invested into the business or distributed to shareholders).

- *Environmental impact*: at a minimum, an inclusive business model does not have major negative environmental impacts and, at best, contributes directly to environmental sustainability (e.g. by saving resources, reducing carbon emissions, conserving biodiversity, etc.).

Additional criteria that can be considered are:

- *Potential for scale* (i.e. expanding regionally, reaching deeper into poor populations, or extending its activities) and replication (by others in the same region/sector).
- *Innovation*: many inclusive business models employ new solutions, being based on an innovative product, process or business model, as well as new ideas that can excite and inspire others.

Many examples of inclusive business models have been documented by the United Nations Development Programme and made available on an online searchable database.

Benefits for Business

- *Generating profits*. Business with the poor can sometimes yield higher rates of return than ventures in developed markets. Some microfinance institutions for instance have demonstrated their ability to reap significant profits.
- *Developing new markets*. The 4 billion people living on less than $8 a day worldwide have a combined income of about $5 trillion. They are willing and able to pay for essential goods and services (such as water, energy and healthcare), but too often they suffer from a 'poverty penalty' and end up paying more than rich consumers. Business models that offer better value for money—or entirely new products and services to improve the lives of the poor—can reap pioneer profits in return.
- *Driving innovation*. The challenge of developing inclusive business models can lead to innovations that contribute to a company's competitiveness. For example, to meet the poor's preferences and needs, firms must offer new combinations of price and performance. And the pervasive constraints that businesses encounter when doing business with the poor require creative responses. These forces drive the development of new products, services and business models that can catch on in other markets, giving innovative companies a competitive advantage in poor markets.
- *Expanding the labour pool*. The poor are a large source of labour. The advantages of hiring them as employees go beyond cost savings. With adequate training and well-targeted marketing, the poor can deliver high-quality products and services. Or their local knowledge and connections may place them well to serve other poor consumers in their communities.

- *Strengthening value chains.* For firms that procure locally, incorporating the poor in business value chains—as producers, suppliers, distributors, retailers and franchisees—can expand supply and lower risk. That allows them to reduce costs and increase flexibility, especially as the local businesses move into more specialized or higher-skill activities such as component production and business services.

Benefits for the Poor

Businesses can improve the lives of poor people, contributing broadly to what the United Nations terms 'human development'—expanding people's opportunities to lead lives they value.

- *Creating jobs and Increasing incomes,* by including poor people in value chains as customers, employees, producers and small-business owners.
- *Meeting basic needs',* such as food, clean water, sanitation, electricity and health-related services all meet people's basic needs.
- *Increasing productivity,* through access to products and services—from electricity to mobile telephony, from agricultural equipment to credit and insurance.
- *Empowering the poor.* All these contributions support the empowerment of poor people, individually and communally, to gain more control over their lives. By raising awareness, by providing information and training, by including marginalized groups, by offering new opportunities and by conferring hope and pride, inclusive business models can give people confidence and new sources of strength to escape poverty using their own means.

As such, inclusive business models can make a significant contribution towards meeting the Millennium Development Goals (MDGs).

Constraints

Although opportunities are great, many businesses are not taking advantage of them because market conditions surrounding the poor can make doing business difficult, risky and expensive. Where poverty prevails, the foundations for functional markets are often lacking, excluding the poor from meaningful participation and deterring companies from doing business with them.

The United Nations Development Programme, in a report titled "Creating Value for All: Strategies for Doing Business with the Poor" (2008), identifies five major market constraints and successful strategies to overcome them:

- *Limited market information.* Businesses know too little about the poor—what poor consumers prefer, what they can afford and what products and capabilities they have to offer as employees, producers and business owners.

- *Ineffective regulatory environments.* The markets of the poor lack regulatory frameworks that allow business to work. Rules and contracts are not enforced. People and enterprises lack access to the opportunities and protections afforded by a functioning legal system.

- *Inadequate physical infrastructure.* Transportation is constrained by the lack of roads and supporting infrastructure. Water, electricity, sanitation and telecommunications networks are lacking.

- *Missing knowledge and skills.* Poor consumers may not know the use and benefits of particular products, or may lack the skills to use them effectively. Poor suppliers, distributors and retailers may lack the knowledge and skills to deliver quality products and services consistently, on time and at a set cost.

- *Restricted access to financial products and financial services.* Lacking credit, poor producers and consumers cannot finance investments or large purchases. Lacking insurance, they cannot protect their meagre assets and income against shocks such as illness, drought or theft. And in the absence of transactional banking services, their financing is insecure and expensive.

Success Factors

Despite these challenges, a growing number of businesses are operating successfully in poor markets. To do so, they use five core strategies:

- *Adapt products and processes.* Information and communications technologies have created the possibility for many such adaptations, including mobile banking (m-banking), smart cards (for instance, to buy water) and telemedicine. M-banking has freed banking processes from relying on brick-and-mortar branches and automated teller machines, infrastructure that rarely exists where poor people live. Customers can now wire money, receive remittances, pay for purchases and service their credit, all through their mobile phones. But businesses are also using other technologies, such as for water purification and off-grid electricity production. In addition, some innovative technological approaches are reducing the use of resources—tying the goal of human development to that of environmental sustainability.

Restructuring business processes can be as important as using new technologies. For example, the global spread of telephony is driven by wireless technology. But bringing mobile telephone service to poor people has depended partly on a change in the business process—the move to selling air time on prepaid cards. With 'smart' payment and pricing methods, an inclusive business model can accommodate the cash flow of its customers and suppliers, who are constrained by low and unreliable incomes and a lack of access to financial services.

- *Invest to remove market constraints.* Investing to remove constraints is cost-effective for business when it creates—or can be made to create—private value that is tangible and capturable, ensuring sufficient benefits to the company. Investing to remove market constraints can create public as well as private value. For example, when a firm educates and trains its employees it creates a more skilled workforce—a shared resource as workers move on to other jobs and companies. This added social value opens up doors for cost-sharing with socially minded funding sources, such as international donors, individual philanthropists, nonprofit social investment funds and governments.

- *Leverage the strengths of the poor.* The poor are often an inclusive business model's most important partners. By engaging the poor as intermediaries and building on their social networks, a company can increase access, trust and accountability. Those qualities in turn help businesses to nurture their markets and expand participation in their value chains. One model for engaging the poor into one's sales operations is microfranchising. Firms can leverage local knowledge and trust by employing the poor to gather market information, to deliver, collect and service products and to train others. Furthermore, poor people often have the best ideas for creating new products and services that meet other poor consumers' needs. Generally, when the poor take over some tasks in a business model, the transaction costs for the business fall—while the poor benefit from rising income, knowledge and skills and social standing.

- *Combine resources and capabilities with others.* Like many business models, inclusive business models often succeed by engaging other businesses in mutually beneficial partnerships and collaborations. They also make use of collaborations with nontraditional partners, such as nongovernmental organizations and public service providers. Through such collaborations, businesses can gain access to complementary capabilities and pool resources to work around or remove constraints in the market environment.

- *Engage in policy dialogue with governments.* Engaging in policy dialogue is an important part of doing business with the poor, where companies are typically first movers and much of the environment for doing business has yet to be built. All market constraints previously mentioned are more or less in the domain of public policy. Policymaking is complex and continual, and businesses can provide good information about the problems and their possible solutions. Sometimes the individual efforts of entrepreneurs and companies to engage with governments can have large implications, such as changing market structures or even opening new markets.

Businesses can also rely on demonstration effects or engage collectively to inform public policy and promote effective regulations in developing countries. Since business engagement in policymaking can be controversial, companies and policymakers need a space to engage in frank yet transparent dialogue about how to improve the business environment. Collaborative efforts can open such a space. Companies operating in the same industry or region often share policy interests. And if they are doing business in ways that contribute to economic opportunity and human development, organizations outside the private sector may have complementary policy interests. Where business models are inclusive, collective action can give businesses a strong and legitimate voice in policymaking.

References

- George,G and Bock AJ. 2012. Models of opportunity: How entrepreneurs design firms to achieve the unexpected. Cambridge University Press, ISBN 978-0-521-17084-0
- Vitasek, Kate. Vested Outsourcing: Five Rules that will Transform Outsourcing" (New York: Palgrave Macmillan, 2012) ISBN 978-1-137-29719-8
- Zuckoff, Mitchell (10 January 2006). Ponzi's scheme - the true story fo a financial legend. New York: Random House Trade Paperbacks. ISBN 0812968360.
- Edwards, Paul (1997). Franchising & licensing: two powerful ways to grow your business in any economy. Tarcher. p. 356. ISBN 0-87477-898-0.
- Clegg, Brian (2000). The invisible customer: strategies for successive customer service down the wire. Kogan Page. p. 112. ISBN 0-7494-3144-X.
- Kitching, Trevor (2001). Purchasing scams and how to avoid them. Gower Publishing Company. p. 4. ISBN 0-566-08281-0.
- Mendelsohn, Martin (2004). The guide to franchising. Cengage Learning Business Press. p. 36. ISBN 1-84480-162-4.
- Blythe, Jim (2004). Sales & Key Account Management. Cengage Learning Business Press. p. 278. ISBN 1-84480-023-7.
- Carroll, Robert Todd (2003). The Skeptic's Dictionary: A Collection of Strange Beliefs, Amusing Deceptions, and Dangerous Delusions. Wiley. p. 235. ISBN 0-471-27242-6.
- Salinger (Editor), Lawrence M. (2005). Encyclopedia of White-Collar & Corporate Crime. 2. Sage Publishing. p. 880. ISBN 0-7619-3004-3.
- Popp, Dr. Karl Michael (2015). Best Practices for commercial use of open source software. Norderstedt, Germany: Books on Demand. ISBN 978-3738619096.
- Solatan, Jean (2011). Advances in software economics: A reader on business models and Partner Ecosystems in the software industry. Norderstedt, Germany: BOD. ISBN 978-3-8448-0405-8.
- "Bountysource Raises $1.1 Million for the First Crowdfunding Platform for Open-Source Software Projects". Yahoo Finance. Marketwired. 16 July 2013. Retrieved 18 June 2016.
- Wheeler, David A. (February 2009). "F/LOSS is Commercial Software". Technology Innovation Management Review. Talent First Network. Retrieved 18 June 2016.
- Germain, Jack M. (5 November 2013). "FOSS in the Enterprise: To Pay or Not to Pay?". LinuxInsider. ECT News Network, Inc. Retrieved 18 June 2016.
- Markham, Gervase (16 March 2004). "Mozilla Foundation Open Letter Orders Unofficial Mozilla Merchandise Sellers to Stop, Legal Action Hinted". MozillaZine. Retrieved 18 June 2016.

- Gartner, Samantha (6 October 2014). “Moodle will always be an open source project”. opensource.com. Retrieved 18 June 2016.
- Dougiamas, Martin (22 January 2014). “Moodle: a case study in sustainability”. OSS Watch. University of Oxford (published 5 June 2007). Retrieved 18 June 2016.
- “How do the Moodle Partners work?”. Moodle. 2012. Archived from the original on 22 July 2014. Retrieved 18 June 2016.
- Kolowich, Steve (27 March 2012). “Blackboard’s Open-Source Pivot”. Inside Higher Ed. Retrieved 18 June 2016.
- Tiemann, Michael (7 June 2007). “GNU Affero GPL version 3 and the “ASP loophole””. Open Source Initiative. Retrieved 18 June 2016.

3

Business Process Modeling: An Overview

Business process modeling is the activity used in systems engineering. The use of this activity is to represent processes of any enterprise. Some aspects of business process modeling elucidated in this section are business reference model, business process and direct selling. This chapter is an overview of the subject matter incorporating all the major aspects of business process modeling.

Business Process Modeling

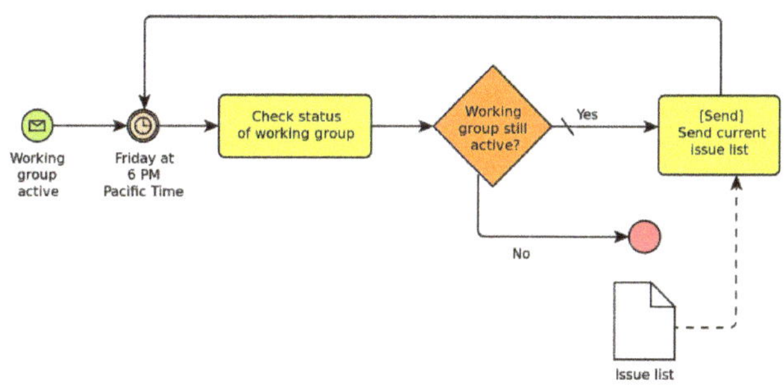

Example of business process modeling of a process with a normal flow with the Business Process Modeling Notation

Business process modeling (BPM) in systems engineering is the activity of representing processes of an enterprise, so that the current process may be analyzed or improved. BPM is typically performed by business analysts, who provide expertise in the modeling discipline; by subject matter experts, who have specialized knowledge of the processes being modeled; or more commonly by a team comprising both. Alternatively, the process model can be derived directly from events' logs using process mining tools.

The business objective is often to increase process speed or reduce cycle time; to increase quality; or to reduce costs, such as labor, materials, scrap, or capital costs. In practice, a management decision to invest in business process modeling is often motivated by the need to document requirements for an information technology project.

Change management programs are typically involved to put any improved business processes into practice. With advances in software design, the vision of BPM models becoming fully executable is coming closer to reality.

History

Techniques to model business process such as the flow chart, functional flow block diagram, control flow diagram, Gantt chart, PERT diagram, and IDEF have emerged since the beginning of the 20th century. The Gantt charts were among the first to arrive around 1899, the flow charts in the 1920s, Functional Flow Block Diagram and PERT in the 1950s, Data Flow Diagrams and IDEF in the 1970s. Among the modern methods are Unified Modeling Language and Business Process Model and Notation. Still, these represent just a fraction of the methodologies used over the years to document business processes. The term 'business process modeling' was coined in the 1960s in the field of systems engineering by S. Williams in his 1967 article 'Business Process Modeling Improves Administrative Control'. His idea was that techniques for obtaining a better understanding of physical control systems could be used in a similar way for business processes. It was not until the 1990s that the term became popular.

In the 1990s the term 'process' became a new productivity paradigm. Companies were encouraged to think in *processes* instead of *functions* and *procedures*. Process thinking looks at the chain of events in the company from purchase to supply, from order retrieval to sales, etc. The traditional modeling tools were developed to illustrate time and cost, while modern tools focus on cross-functional activities. These cross-functional activities have increased significantly in number and importance, due to the growth of complexity and dependence. New methodologies include business process redesign, business process innovation, business process management, integrated business planning, among others, all "aiming at improving processes across the traditional functions that comprise a company".

In the field of software engineering, the term 'business process modeling' opposed the common software process modeling, aiming to focus more on the state of the practice during software development. In that time (early 1990s) all existing and new modeling techniques to illustrate business processes were consolidated as 'business process modeling languages'. In the Object Oriented approach, it was considered to be an essential step in the specification of business application systems. Business process modeling became the base of new methodologies, for instance those that supported data collection, data flow analysis, process flow diagrams and reporting facilities. Around 1995, the first visually oriented tools for business process modeling and implementation were being presented.

Topics

Business Model

A business model is a framework for creating economic, social, and/or other forms of value. The term 'business model' is thus used for a broad range of informal and formal descriptions to represent core aspects of a business, including purpose, offerings, strategies, infrastructure, organizational structures, trading practices, and operational processes and policies.

In the most basic sense, a business model is the method of doing business by which a company can sustain itself. That is, generate revenue. The business model spells-out how a company makes money by specifying where it is positioned in the value chain.

Business Process

A business process is a collection of related, structured activities or tasks that produce a specific service or product (serve a particular goal) for a particular customer or customers. There are three main types of business processes:

1. Management processes, that govern the operation of a system. Typical management processes include corporate governance and strategic management.
2. Operational processes, that constitute the core business and create the primary value stream. Typical operational processes are purchasing, manufacturing, marketing, and sales.
3. Supporting processes, that support the core processes. Examples include accounting, recruitment, and technical support.

A business process can be decomposed into several sub-processes, which have their own attributes, but also contribute to achieving the goal of the super-process. The analysis of business processes typically includes the mapping of processes and sub-processes down to activity level. A business process model is a model of one or more business processes, and defines the ways in which operations are carried out to accomplish the intended objectives of an organization. Such a model remains an abstraction and depends on the intended use of the model. It can describe the workflow or the integration between business processes. It can be constructed in multiple levels.

A workflow is a depiction of a sequence of operations, declared as work of a person, of a simple or complex mechanism, of a group of persons, of an organization of staff, or of machines. Workflow may be seen as any abstraction of real work, segregated into workshare, work split or other types of ordering. For control purposes, workflow may be a view of real work under a chosen aspect.

Artifact-centric Business Process

The artifact-centric business process model has emerged as a holistic approach for modeling business processes, as it provides a highly flexible solution to capture operational specifications of business processes. It particularly focuses on describing the data of business processes, known as "artifacts", by characterizing business-relevant data objects, their lifecycles, and related services. The artifact-centric process modelling approach fosters the automation of the business operations and supports the flexibility of the workflow enactment and evolution.

Tools

Business process modeling tools provide business users with the ability to model their business processes, implement and execute those models, and refine the models based on as-executed data. As a result, business process modeling tools can provide transparency into business processes, as well as the centralization of corporate business process models and execution metrics.

Modeling and Simulation

Modeling and simulation functionality allows for pre-execution "what-if" modeling and simulation. Post-execution optimization is available based on the analysis of actual as-performed metrics.

- Use case diagrams created by Ivar Jacobson, 1992 (integrated in UML)
- Activity diagrams (also adopted by UML)

Some business process modeling techniques are:

- Business Process Model and Notation (BPMN)
- Cognition enhanced Natural language Information Analysis Method (CogNIAM)
- Extended Business Modeling Language (xBML)
- Event-driven process chain (EPC)
- ICAM DEFinition (IDEF0)
- Unified Modeling Language (UML), extensions for business process such as Eriksson-Penker's
- Formalized Administrative Notation (FAN)

Programming Language Tools

BPM suite software provides programming interfaces (web services, application program interfaces (APIs)) which allow enterprise applications to be built to leverage the BPM engine. This component is often referenced as the *engine* of the BPM suite.

Programming languages that are being introduced for BPM include:

- Business Process Execution Language (BPEL),
- Web Services Choreography Description Language (WS-CDL).
- XML Process Definition Language (XPDL),

Some vendor-specific languages:

- Architecture of Integrated Information Systems (ARIS) supports EPC,
- Java Process Definition Language (JBPM),

Other technologies related to business process modeling include model-driven architecture and service-oriented architecture.

Business Reference Model

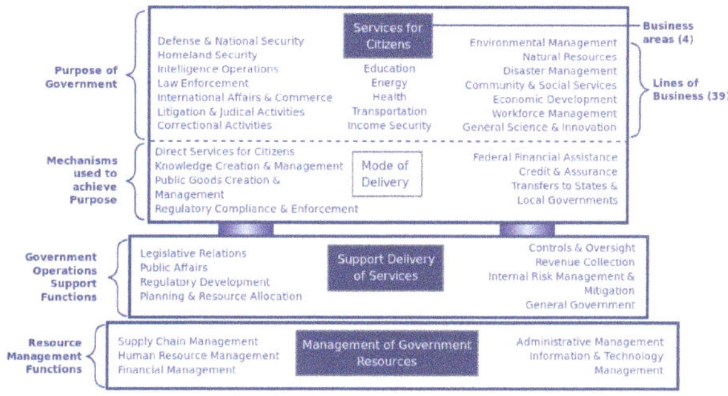

Example of the US Federal Government Business Reference Model

A business reference model is a reference model, concentrating on the functional and organizational aspects of an enterprise, service organization or government agency. In general a reference model is a model of something that embodies the basic goal or idea of something and can then be looked at as a reference for various purposes. A business reference model is a means to describe the business operations of an organization, independent of the organizational structure that perform them. Other types of business reference model can also depict the relationship between the business processes, business functions, and the business area's business reference model. These reference models can be constructed in layers, and offer a foundation for the analysis of service components, technology, data, and performance.

The most familiar business reference model is the Business Reference Model of the US federal government. That model is a function-driven framework for describing the business operations of the federal government independent of the agencies that perform them. The Business Reference Model provides an organized, hierarchical construct for describing the day-to-day business operations of the federal government. While many models exist for describing organizations – organizational charts, location maps, etc. – this model presents the business using a functionally driven approach.

Business Process Integration

A business model, which may be considered an elaboration of a business process model, typically shows business data and business organizations as well as business processes. By showing business processes and their information flows a business model allows business stakeholders to define, understand, and validate their business enterprise. The data model part of the business model shows how business information is stored, which is useful for developing software code. See the figure on the right for an example of the interaction between business process models and data models.

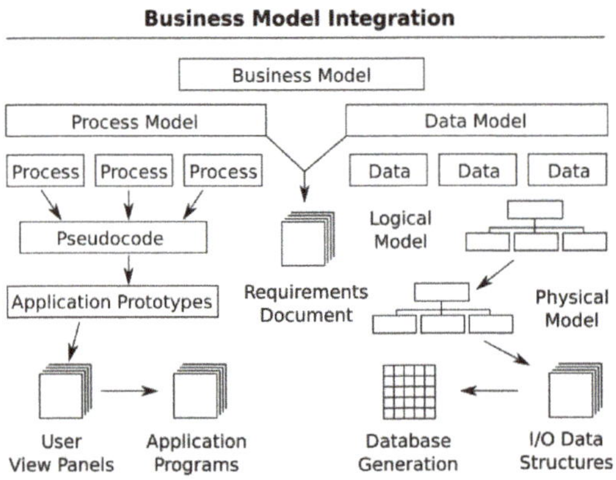

Example of the interaction between business process and data models

Usually a business model is created after conducting an interview, which is part of the business analysis process. The interview consists of a facilitator asking a series of questions to extract information about the subject business process. The interviewer is referred to as a facilitator to emphasize that it is the participants, not the facilitator, who provide the business process information. Although the facilitator should have some knowledge of the subject business process, but this is not as important as the mastery of a pragmatic and rigorous method interviewing business experts. The method is important because for most enterprises a team of facilitators is needed to collect information across the enterprise, and the findings of all the interviewers must be compiled and integrated once completed.

Business models are developed as defining either the current state of the process, in which case the final product is called the "as is" snapshot model, or a concept of what the process should become, resulting in a "to be" model. By comparing and contrasting "as is" and "to be" models the business analysts can determine if the existing business processes and information systems are sound and only need minor modifications, or if reengineering is required to correct problems or improve efficiency. Consequently, business process modeling and subsequent analysis can be used to fundamentally reshape the way an enterprise conducts its operations.

Business Process Reengineering

Business process reengineering (BPR) aims to improve the efficiency and effectiveness of the processes that exist within and across organizations. It examines business processes from a "clean slate" perspective to determine how best to construct them.

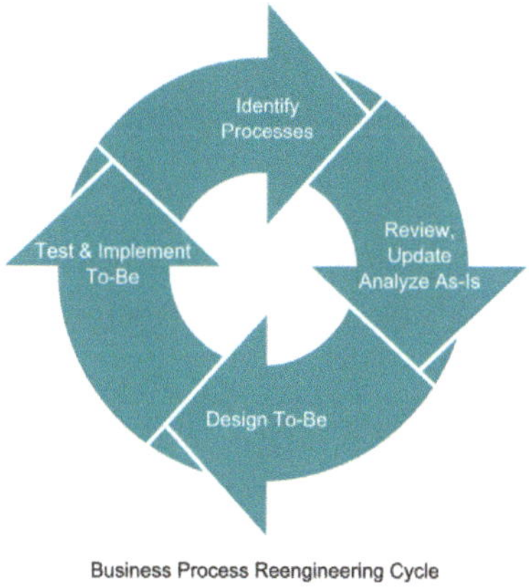

Business Process Reengineering Cycle

Business process reengineering (BPR) began as a private sector technique to help organizations fundamentally rethink how they do their work. A key stimulus for reengineering has been the development and deployment of sophisticated information systems and networks. Leading organizations use this technology to support innovative business processes, rather than refining current ways of doing work. yogesh.

Business Process Management

Business process management is a field of management focused on aligning organizations with the wants and needs of clients. It is a holistic management approach that promotes business effectiveness and efficiency while striving for innovation, flexibility and integration with technology. As organizations strive for attainment of their objectives, business process management attempts to continuously improve processes - the process to define, measure and improve your processes – a "process optimization" process.

An Organizational Improvement Methodology focused on aligning and driving strategy through operations and leveraging information requirements.

Business Reference Model

Business reference model (BRM) is a reference model, concentrating on the functional and organizational aspects of the core business of an enterprise, service organization or government agency.

In enterprise engineering a business reference model is part of an Enterprise Architecture Framework or *Architecture Framework*. An Enterprise Architecture Framework defines in a series of reference models, how to organize the structure and views associated with an Enterprise Architecture.

Overview

A reference model in general is a model of something that embodies the basic goal or idea of something and can then be looked at as a reference for various purposes. A business reference model is a means to describe the business operations of an organization, independent of the organizational structure that perform them. Other types of business reference model can also depict the relationship between the business processes, business functions, and the business area's business reference model. These reference model can be constructed in layers, and offer a foundation for the analysis of service components, technology, data, and performance.

The most familiar business reference model is the "Business Reference Model", one of five reference models of the Federal Enterprise Architecture of the US Federal Government. That model is a function-driven framework for describing the business operations of the Federal Government independent of the agencies that perform them. The Business Reference Model provides an organized, hierarchical construct for describing the day-to-day business operations of the Federal government. While many models exist for describing organizations - organizational charts, location maps, etc. - this model presents the business using a functionally driven approach.

History

One of the first business reference models ever defined was the "IMPPACT Business Reference Model" around 1990, which was the result of a research project in the Computer Integrated Manufacturing (CIM) field of the ESPRIT1 programme. Gielingh et al. (1933) described:

The IMPPACT Business Reference Model is expressed in the generic language constructs provided by IDEF0... It describes the requirements for CIM seen from a business point of view. Views modelled are manufacturing activities, real and information flow objects resource objects (information and material processing components) and organisational aspects (departments and their relations to activities and resources). The complete manufacturing system (including the production system and its management) is modelled by the IMPPACT Business Reference Model. Management covers both the planning of the production and the planning and control of this production.

The term IMPPACT stood for Integrated Manufacturing of Products and Processes using Advanced Computer Technologies Furthermore, in its framework were incorporated CIMOSA as reference model, NIAM for information modelling, and the data modeling language EXPRESS for information structure implementation.

In the 1990s, business reference models were hardly an item. An exception was a 1991 book about IT management, which mentioned that the Kodak management had developed a business reference model 10 years earlier. A 1996 manual of the SAP R/3 enterprise resource planning software stipulated the existence on the business reference model of the R/3 System. However, in the 1990s there was a significant development of reference models in related fields, which, resulted in the developments of Integrated business planning, the Open System Environment Reference Model, the Workflow Reference Model, TOGAF and the Zachman Framework.

In the new millennium business reference models started emerging in several fields from network management systems, and E-business, to the US Federal government. The US Federal government published it's "Business Reference Model", Version 1.0 in February 2002. Related developments in this decade were the development of the Treasury Enterprise Architecture Framework, and the OASIS SOA Reference Model.

Specific Models

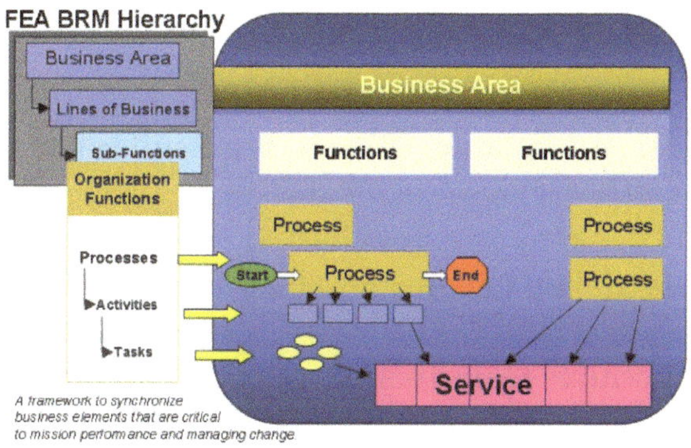

Other view on the FEA Business reference model.

The US Federal Government has defined a Federal Enterprise Architecture structures of the five FEA reference models:

- Performance Reference Model (PRM)
- *Business Reference Model* (BRM)
- Service Component Reference Model (SRM)
- Technical Reference Model (TRM)
- Data Reference Model (DRM)

The Federal Government Business Reference Model (FA BRM) provides an organized, hierarchical construct for describing the day-to-day business operations of the Federal government. While many models exist for describing organizations - org charts, location maps, etc. - this model presents the business using a functionally driven approach. The Lines of Business and Sub-functions that comprise the BRM represent a departure from previous models of the Federal government that use antiquated, stovepiped, agency-oriented frameworks. The BRM is the first layer of the Federal Enterprise Architecture and it is the main viewpoint for the analysis of data, service components and technology.

Business Process Model and Notation

Business Process Model and Notation (BPMN) is a graphical representation for specifying business processes in a business process model.

Business Process Management Initiative (BPMI) developed BPMN, which has been maintained by the Object Management Group since the two organizations merged in 2005. Version 2.0 of BPMN was released in January 2011, at which point the name was adapted to Business Process Model and Notation as execution semantics were also introduced alongside the notational and diagramming elements.

Overview

Business Process Model and Notation (BPMN) is a standard for business process modeling that provides a graphical notation for specifying business processes in a *Business Process Diagram* (BPD), based on a flowcharting technique very similar to activity diagrams from Unified Modeling Language (UML). The objective of BPMN is to support business process management, for both technical users and business users, by providing a notation that is intuitive to business users, yet able to represent complex process semantics. The BPMN specification also provides a mapping between the graphics of the notation and the underlying constructs of execution languages, particularly Business Process Execution Language (BPEL).

The primary goal of BPMN is to provide a standard notation readily understandable by all business stakeholders. These include the business analysts who create and refine the processes, the technical developers responsible for implementing them, and the business managers who monitor and manage them. Consequently, BPMN serves as a common language, bridging the communication gap that frequently occurs between business process design and implementation.

Currently there are several competing standards for business process modeling languages used by modeling tools and processes. Widespread adoption of a single standard would help unify the expression of basic business process concepts (e.g., public and private processes, choreographies), as well as advanced process concepts (e.g., exception handling, transaction compensation).

Since 2014, BPMN has been complemented by a new standard for building decision models, the Decision Model and Notation standard.

BPMN Topics

Scope

BPMN is constrained to support only the concepts of modeling applicable to business processes. Other types of modeling done by organizations for non-process purposes are out of scope for BPMN. Examples of modeling excluded from BPMN are:

- Organizational structures
- Functional breakdowns
- Data models

In addition, while BPMN shows the flow of data (messages), and the association of data artifacts to activities, it is not a data flow diagram.

Elements

BPMN models consist of simple diagrams constructed from a limited set of graphical elements. For both business users and developers, they simplify understanding business activities' flow and process. BPMN's four basic element categories are:

Flow objects

Events, activities, gateways

Connecting objects

Sequence flow, message flow, association

Swim lanes

Pool, lane

Artifacts

Data object, group, annotation

These four categories enable creation of simple business process diagrams (BPDs). BPDs also permit making new types of flow object or artifact, to make the diagram more understandable.

Flow Objects and Connecting Objects

Event

Activity

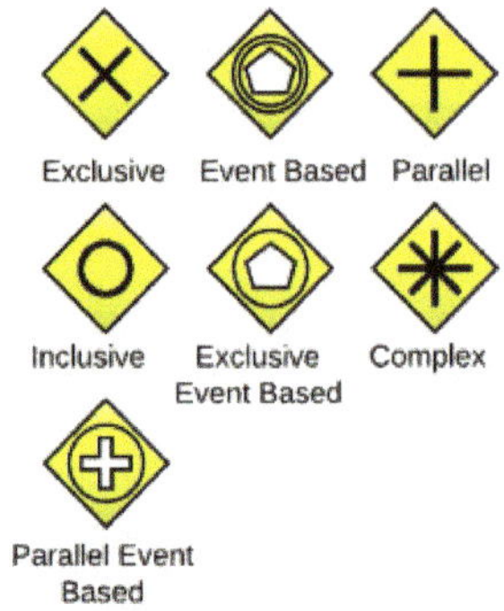

Gateway

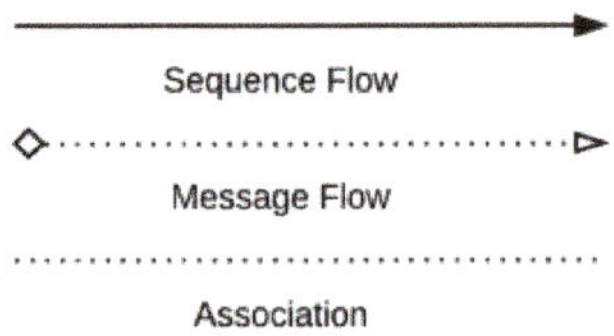

Connections

Flow objects are the main describing elements within BPMN, and consist of three core elements: events, activities, and gateways.

Event

An Event is represented with a circle and denotes something that *happens* (compared with an activity, which is something that is *done*). Icons within the circle denote the type of event (e.g., an envelope representing a message, or a clock representing time). Events are also classified as Catching (for example, if catching an incoming message starts a process) or Throwing (such as throwing a completion message when a process ends).

Start event

Acts as a process trigger; indicated by a single narrow border, and can only be *Catch*, so is shown with an open (outline) icon.

Intermediate event

Represents something that happens between the start and end events; is indicated by a double border, and can *Throw* or *Catch* (using solid or open icons as appropriate). For example, a task could flow to an event that throws a message across to another pool, where a subsequent event waits to catch the response before continuing.

End event

Represents the result of a process; indicated by a single thick or bold border, and can only *Throw*, so is shown with a solid icon.

Activity

An activity is represented with a rounded-corner rectangle and describes the kind of work which must be done. An activity is a generic term for work that a company performs. It can be atomic or compound.

Task

A task represents a single unit of work that is not or cannot be broken down to a further level of business process detail. It is referred to as an atomic activity. A task is the lowest level activity illustrated on a process diagram. A set of tasks may represent a high-level procedure.

Sub-process

Used to hide or reveal additional levels of business process detail. When collapsed, a sub-process is indicated by a plus sign against the bottom line of the rectangle; when expanded, the rounded rectangle expands to show all flow objects, connecting objects, and artifacts. A sub-process is referred to as a compound activity.

Has its own self-contained start and end events; sequence flows from the *parent* process must not cross the boundary.

Transaction

A form of sub-process in which all contained activities must be treated as a whole; i.e., they must all be completed to meet an objective, and if any one of them fails, they must all be compensated (undone). Transactions are differentiated from expanded sub-processes by being surrounded by a double border.

Call Activity

A point in the process where a global process or a global Task is reused. A call activity is differentiated from other activity types by a bolded border around the activity area.

Gateway

A gateway is represented with a diamond shape and determines forking and merging of paths, depending on the conditions expressed.

Exclusive

Used to create alternative flows in a process. Because only one of the paths can be taken, it is called exclusive.

Event Based

The condition determining the path of a process is based on an evaluated event.

Parallel

Used to create parallel paths without evaluating any conditions.

Inclusive

Used to create alternative flows where all paths are evaluated.

Exclusive Event Based

An event is being evaluated to determine which of mutually exclusive paths will be taken.

Complex

Used to model complex synchronization behavior.

Parallel Event Based

Two parallel processes are started based on an event, but there is no evaluation of the event.

Connections

Flow objects are connected to each other using Connecting objects, which are of three types: sequences, messages, and associations.

Sequence Flow

A Sequence Flow is represented with a solid line and arrowhead, and shows in which order the activities are performed. The *sequence flow* may also have a symbol at its start, a small diamond indicates one of a number of conditional flows from an activity, while a diagonal slash indicates the default flow from a decision or activity with conditional flows.

Message Flow

A Message Flow is represented with a dashed line, an open circle at the start, and an open arrowhead at the end. It tells us what messages flow across organizational boundaries (i.e., between pools). A message flow can never be used to connect activities or events within the same pool.

Association

An Association is represented with a dotted line. It is used to associate an Artifact or text to a Flow Object, and can indicate some directionality using an open arrowhead (toward the artifact to represent a result, from the artifact to represent an input, and both to indicate it is read and updated). No directionality is used when the Artifact or text is associated with a sequence or message flow (as that flow already shows the direction).

Swim Lanes and Artifacts

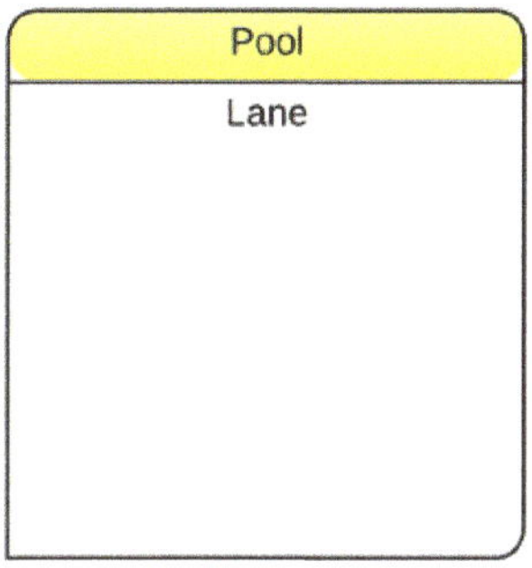

Swimlanes

Data objects

Groups

Annotation

Swim lanes are a visual mechanism of organising and categorising activities, based on cross functional flowcharting, and in BPMN consist of two types:

Pool

Represents major participants in a process, typically separating different organisations. A pool contains one or more lanes (like a real swimming pool). A pool can be open (i.e., showing internal detail) when it is depicted as a large rectangle showing one or more lanes, or collapsed (i.e., hiding internal detail) when it is depicted as an empty rectangle stretching the width or height of the diagram.

Lane

Used to organise and categorise activities within a pool according to function or role, and depicted as a rectangle stretching the width or height of the pool. A lane contains the flow objects, connecting objects and artifacts.

Artifacts allow developers to bring some more information into the model/diagram. In this way the model/diagram becomes more readable. There are three pre-defined Artifacts and they are:

- Data objects: Data objects show the reader which data is required or produced in an activity.
- Group: A Group is represented with a rounded-corner rectangle and dashed lines. The group is used to group different activities but does not affect the flow in the diagram.
- Annotation: An annotation is used to give the reader of the model/diagram an understandable impression.

Examples of Business Process Diagrams

- Click on small images for full-size version

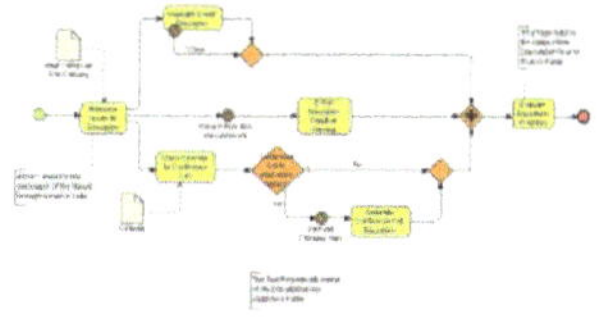

Discussion cycle

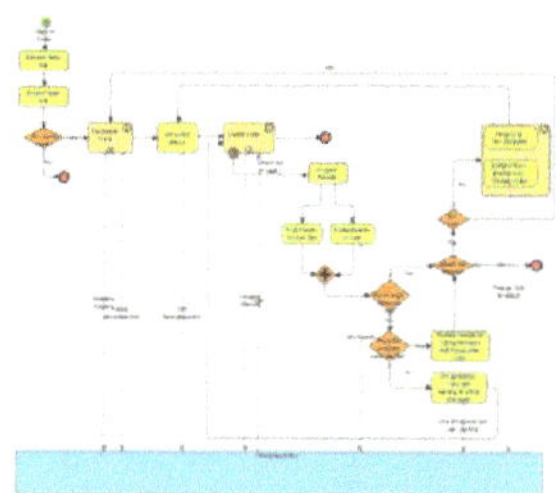

E-mail voting process

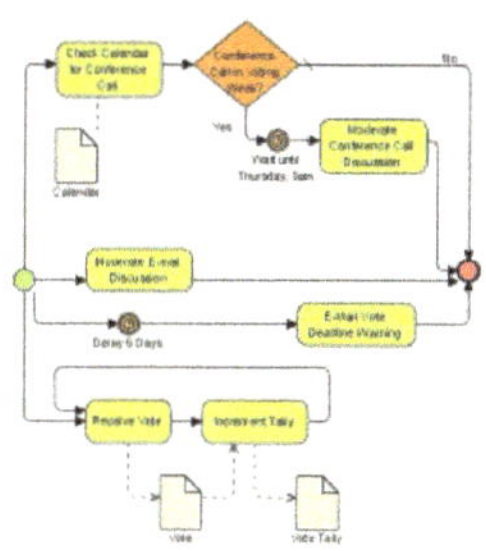

Collect votes

BPMN 2.0

The vision of BPMN 2.0 is to have one single specification for a new Business Process Model and Notation that defines the notation, metamodel and interchange format but with a modified name that still preserves the "BPMN" brand. The features include

- Aligning BPMN with the business process definition meta model BPDM to form a single consistent language.
- Enabling the exchange of business process models and their diagram layouts among process modeling tools to preserve semantic integrity.
- Expand BPMN to allow model orchestrations and choreographies as stand-alone or integrated models.
- Support the display and interchange of different perspectives on a model that allow a user to focus on specific concerns.
- Serialize BPMN and provide XML schemes for model transformation and to extend BPMN towards business modeling and executive decision support.

The final version of the specification was released in January, 2011.

Comparison of BPMN Versions

Attributes	BPMN 1.0	BPMN 1.1	BPMN 1.2	BPMN 2.0
Consortium	BPMI	OMG	OMG	OMG
Date of release	May 2004	January 2008	January 2009	January 2011
Models	• Collaborative (public) B2B processes, • internal (private) business processes.			• collaborative (public) B2B processes, • internal (private) business processes, • a choreography – expected behavior between two or more business participants, • collaborations, which is a collection of participants and their interaction and • a conversation – the logical relation of message exchanges.
event	• **start** (none, message, timer, rule, link, multiple) • **intermediate** (none, message, timer, error, cancel, compensation, rule, link, multiple) • **end** (none, message, error, cancel, compensation, link, terminate, multiple)	• **start** (none, message, timer, conditional, signal, multiple) • **intermediate** (none, message, timer, error, cancel, compensation, conditional, link, signal, multiple) • **end** (none, message, error, cancel, compensation, signal,terminate, multiple)		• **start** ○ **top-level** (none, message, timer, conditional, signal, multiple, parallel multiple) ○ **event sub-process interrupting** (message, timer, escalation, conditional, error, compensation, signal, multiple, parallel multiple) ○ **event sub-process non-interrupting** (message, timer, escalation, conditional, signal, multiple, parallel multiple) • **intermediate** ○ **catching** (message, timer, conditional, link, signal, multiple, parallel multiple) ○ **boundary interrupting** (message, timer, escalation, conditional, error, cancel, compensation, signal, multiple, parallel multiple) ○ **boundary non-interrupting** (message, timer, escalation, conditional, signal, multiple, parallel multiple, terminate) ○ **throwing** (none, message, escalation, link, compensation, signal, multiple, parallel multiple) • **end** (none, message, escalation, error, cancel, compensation, signal, multiple, terminate)

activity	• **task** (atomic) • **process/sub-process** (nonatomic) ○ collapsed sub-process ○ expanded sub-process			• **task** (atomic) • **choreography task** ○ collapsed choreography sub-process ○ expanded choreography sub-process • **process/sub-process** (nonatomic) ○ collapsed sub-process ○ expanded sub-process
gateway	• **XOR** – exclusive decision and merging. both data-based and event-based. data-based can be shown with or without the "x" marker. • **OR** – inclusive decision and merging • **complex** – complex conditions and situations • **AND** – forking and joining	• **exclusive** decision and merging. both data-based and event-based. data-based can be shown with or without the "x" marker. • **inclusive** decision and merging. • **complex** – complex conditions and situations. • **parallel** forking and joining.		• **exclusive** decision and merging. both data-based and event-based. exclusive can be shown with or without the "x" marker. • **inclusive** gateway decision and merging • **complex gateway** – complex conditions and situations • **parallel gateway** – forking and joining
sequence flow	**normal flow** uncontrolled flow conditional flow default flow **exception flow**			
message flow	message flow			
association	association			
pool	pool			
lane	lane			
data objects	data object			• data object ○ collection ○ data input ○ data output
groups	group			
annotations	annotations			
message	/	/	/	**message**
other elements	• looping ○ activity looping ○ sequence flow looping • multiple instances • process break • transactions • nested/embedded sub-process • off-page connector • compensation association			• looping ○ activity looping ○ sequence flow looping • multiple instances • process break • transactions • nested/embedded sub-process • off-page connector • compensation association • communication (subcommunication) • communication link
Number of all elements	48	55	55	116

Major changes	/	• The new specification introduces a categorization of event triggers into "catching" and "throwing" events. I.e. there are two kinds of intermediate message events now – one kind responsible for reception of messages ("catching") and one kind responsible for sending messages ("throwing"). • In addition to the old types, it introduces a new type, the **signal event**. • Start and end link events do not exist any longer in BPMN 1.1. • The old "rule events" where renamed to **conditional events**. The semantics and appearance have not changed. • The event-based gateway in BPMN 1.1 looks slightly different from what it looked like in 1.0. Instead of the hexagonal star it now has a pentagon in its center. The same shape is also used for the multiple events (start, intermediate, end). • There is an additional line separating your lane's description from its content.	The BPMN 1.2 minor revision changes consist of editorial corrections and implementation bug fixes. Consequently, these minor changes affect modeling tool vendors more than modelers (users).	• Choreographies o Choreographies-model o Conversation-model • Complete Metamodel • BPMN Core • BPMN Execution Semantics • BPMN – BPEL Mapping • XPDL (BPMN XML Serialization) • Diagram Interchange • Elements For Abstraction o Callable Element o Call Activity o Global Task • Gateways (Updated) o Exclusive/Parallel Event-based Gateway (they stand at the beginning of the process) • Tasks/SubProcesses (Updated) o Event-Subprocess (Used to handle events in the bounding subprocess) o BusinessRule task o Sequential Multi-Instance Activity o Service Task • Artifacts (Updated) o Data Objects (Collection, Data Input, Data Output)

Types of BPMN Sub-model

Business process modeling is used to communicate a wide variety of information to a wide variety of audiences. BPMN is designed to cover this wide range of usage and allows modeling of end-to-end business processes to allow the viewer of the Diagram to be able to easily differentiate between sections of a BPMN Diagram. There are three basic types of sub-models within an end-to-end BPMN model: Private (internal) business processes, Abstract (public) processes, and Collaboration (global) processes:

Private (internal) business processes

> Private business processes are those internal to a specific organization and are the type of processes that have been generally called workflow or BPM processes. If swim lanes are used then a private business process will be contained within a single Pool. The Sequence Flow of the Process is therefore contained within the Pool and cannot cross the boundaries of the Pool. Message Flow can cross the Pool boundary to show the interactions that exist between separate private business processes.

Abstract (public) processes

> This represents the interactions between a private business process and another process or participant. Only those activities that communicate outside the private business process are included in the abstract process. All other "internal" activities of the private business process are not shown in the abstract process. Thus, the abstract process shows to the outside world the sequence of messages that are required to interact with that business process. Abstract processes are contained within a Pool and can be modeled separately or within a larger BPMN Diagram to show the Message Flow between the abstract process activities and other entities. If the abstract process is in the same Diagram as its corresponding private business process, then the activities that are common to both processes can be associated.

Collaboration (global) processes

> A collaboration process depicts the interactions between two or more business entities. These interactions are defined as a sequence of activities that represent the message exchange patterns between the entities involved. Collaboration processes may be contained within a Pool and the different participant business interactions are shown as Lanes within the Pool. In this situation, each Lane would represent two participants and a direction of travel between them. They may also be shown as two or more Abstract Processes interacting through Message Flow (as described in the previous section). These processes can be modeled separately or within a larger BPMN Diagram to show the Associations between the collaboration process activities and other entities. If the collaboration process is in the

same Diagram as one of its corresponding private business process, then the activities that are common to both processes can be associated.

Within and between these three BPMN sub-models, many types of Diagrams can be created. The following are the types of business processes that can be modeled with BPMN (those with asterisks may not map to an executable language):

- High-level private process activities (not functional breakdown)*
- Detailed private business process
- As-is or old business process*
- To-be or new business process
- Detailed private business process with interactions to one or more external entities (or "Black Box" processes)
- Two or more detailed private business processes interacting
- Detailed private business process relationship to Abstract Process
- Detailed private business process relationship to Collaboration Process
- Two or more Abstract Processes*
- Abstract Process relationship to Collaboration Process*
- Collaboration Process only (e.g., ebXML BPSS or RosettaNet)*
- Two or more detailed private business processes interacting through their Abstract Processes and/or a Collaboration Process

BPMN is designed to allow all the above types of Diagrams. However, it should be cautioned that if too many types of sub-models are combined, such as three or more private processes with message flow between each of them, then the Diagram may become too hard for someone to understand. Thus, we recommend that the modeler pick a focused purpose for the BPD, such as a private process, or a collaboration process.

Weaknesses of BPMN

The weaknesses of BPMN could relate to:

- ambiguity and confusion in sharing BPMN models
- support for routine work
- support for knowledge work, and

- converting BPMN models to executable environments
- support for business rules and decision-making

BPEL and BPMN

The BPMN specification includes an informal and partial mapping from BPMN to BPEL 1.1. A more detailed mapping of BPMN to BPEL has been implemented in a number of tools, including an open-source tool known as BPMN2BPEL. However, the development of these tools has exposed fundamental differences between BPMN and BPEL, which make it very difficult, and in some cases impossible, to generate human-readable BPEL code from BPMN models. Even more difficult is the problem of BPMN-to-BPEL *round-trip* engineering: generating BPEL code from BPMN diagrams and maintaining the original BPMN model and the generated BPEL code synchronized, in the sense that any modification to one is propagated to the other.

Business Process

A business process or business method is a collection of related, structured activities or tasks that produce a specific service or product (serve a particular goal) for a particular customer or customers. It may often be visualized as a flowchart of a sequence of activities with interleaving decision points or as a Process Matrix of a sequence of activities with relevance rules based on data in the process.

Overview

There are three types of business processes citation by Prof August-Wilhelm Scheer and Prof. Mark von Rosing:

1. Management processes, the processes that govern the operation of a system. Typical management processes include "corporate governance" and "strategic management".
2. Operational processes, processes that constitute the core business and create the primary value stream. For example, taking orders from customers, and opening an account in a bank branch.
3. Supporting processes, which support the core processes. Examples include Health & Safety, accounting, recruitment, call center, technical support.

A business process begins with a mission objective and ends with achievement of the business objective. Process-oriented organizations break down the barriers of structural departments and try to avoid functional silos.

A complex business process may be decomposed into several sub-processes, which have their own attributes, but also contribute to achieving the goal of the super-process. The analysis of business processes typically includes the mapping of processes and sub-processes down to activity/task level.

Business processes are designed to add value for the customer and should not include unnecessary activities. The outcome of a well designed business process is increased effectiveness (value for the customer) and increased efficiency (less use of resources).

Business Processes can be modeled through a large number of methods and techniques. For instance, the Business Process Modeling Notation is a Business Process Modeling technique that can be used for drawing business processes in a workflow.

History

Adam Smith

An important early (1776) description of processes was that of economist Adam Smith in his famous example of a pin factory. Inspired by an article in Diderot's Encyclopédie, Smith described the production of a pin in the following way:

"One man draws out the wire, another straights it, a third cuts it, a fourth points it, a fifth grinds it at the top for receiving the head: to make the head requires two or three distinct operations: to put it on is a particular business, to whiten the pins is another ... and the important business of making a pin is, in this manner, divided into about eighteen distinct operations, which in some manufactories are all performed by distinct hands, though in others the same man will sometime perform two or three of them."

Smith also first recognized how the output could be increased through the use of labor division. Previously, in a society where production was dominated by handcrafted goods, one man would perform all the activities required during the production process, while Smith described how the work was divided into a set of simple tasks, which would be performed by specialized workers. The result of labor division in Smith's example resulted in productivity increasing by 24,000 percent (sic), i.e. that the same number of workers made 240 times as many pins as they had been producing before the introduction of labor division.

It is worth noting that Smith did not advocate labor division at any price and *per se*. The appropriate level of task division was defined through experimental design of the production process. In contrast to Smith's view which was limited to the same functional domain and comprised activities that are in direct sequence in the manufacturing process, today's process concept includes cross-functionality as an important characteristic. Following his ideas the division of labor was adopted widely, while the integration of tasks into a functional, or cross-functional, process was not considered as an alternative option until much later.

Frederick Winslow Taylor

American engineer, Frederick Winslow Taylor greatly influenced and improved the quality of industrial processes in the early twentieth century. His Principles of Scientific Management focused on standardization of processes, systematic training and clearly defining the roles of management and employees. His methods were widely adopted in the United States, Russia and parts of Europe and led to further developments such as "time and motion study" and visual task optimization techniques, such as Gantt charts.

Peter Drucker

In the latter part of the twentieth century, management guru Peter Drucker focused much of his work on simplification and decentralization of processes, which led to the concept of outsourcing.

Other Definitions

In the early 1990s, US corporations, and subsequently companies all over the world, started to adopt the concept of business process reengineering (BPR) in an attempt to re-achieve the competitiveness that they had lost during the previous decade.

Davenport (1993) defines a (business) process as:

"a structured, measured set of activities designed to produce a specific output for a particular customer or market. It implies a strong emphasis on how work is done within an organization, in contrast to a product focus's emphasis on what. A process is thus a specific ordering of work activities across time and space, with a beginning and an end, and clearly defined inputs and outputs: a structure for action. ... Taking a process approach implies adopting the customer's point of view. Processes are the structure by which an organization does what is necessary to produce value for its customers."

This definition contains certain characteristics a process must possess. These characteristics are achieved by a focus on the business logic of the process (how work is done), instead of taking a product perspective (what is done). Following Davenport's definition of a process we can conclude that a process must have clearly defined boundaries, input and output, that it consists of smaller parts, activities, which are ordered in time and space, that there must be a receiver of the process outcome- a customer - and that the transformation taking place within the process must add customer value.

Hammer & Champy's (1993) definition can be considered as a subset of Davenport's. They define a process as:

"a collection of activities that takes one or more kinds of input and creates an output that is of value to the customer."

As we can note, Hammer & Champy have a more transformation oriented perception, and put less emphasis on the structural component – process boundaries and the order of activities in time and space.

Rummler & Brache (1995) use a definition that clearly encompasses a focus on the organization's external customers, when stating that

"a business process is a series of steps designed to produce a product or service. Most processes (...) are cross-functional, spanning the 'white space' between the boxes on the organization chart. Some processes result in a product or service that is received by an organization's external customer. We call these primary processes. Other processes produce products that are invisible to the external customer but essential to the effective management of the business. We call these support processes."

The above definition distinguishes two types of processes, primary and support processes, depending on whether a process is directly involved in the creation of customer value, or concerned with the organization's internal activities. In this sense, Rummler and Brache's definition follows Porter's value chain model, which also builds on a division of primary and secondary activities. According to Rummler and Brache, a typical characteristic of a successful process-based organization is the absence of secondary activities in the primary value flow that is created in the customer oriented primary processes. The characteristic of processes as spanning the white space on the organization chart indicates that processes are embedded in some form of organizational structure. Also, a process can be cross-functional, i.e. it ranges over several business functions.

Johansson et al. (1993). define a process as:

"a set of linked activities that take an input and transform it to create an output. Ideally, the transformation that occurs in the process should add value to the input and create an output that is more useful and effective to the recipient either upstream or downstream."

This definition also emphasizes the constitution of links between activities and the transformation that takes place within the process. Johansson et al. also include the upstream part of the value chain as a possible recipient of the process output. Summarizing the four definitions above, we can compile the following list of characteristics for a business process:

1. *Definability* : It must have clearly defined boundaries, input and output.
2. *Order* : It must consist of activities that are ordered according to their position in time and space (a sequence).
3. *Customer* : There must be a recipient of the process› outcome, a customer.
4. *Value-adding* : The transformation taking place within the process must add value to the recipient, either upstream or downstream.

5. *Embeddedness* : A process cannot exist in itself, it must be embedded in an organizational structure.
6. *Cross-functionality* : A process regularly can, but not necessarily must, span several functions.

Frequently, identifying a process owner, (i.e., the person responsible for the continuous improvement of the process) is considered as a prerequisite. Sometimes the process owner is the same person who is performing the process.

Related Concepts

Workflow

Workflow is the movement of information or material from one activity or worksite to another. Workflow includes the procedures, people and tools involved in each step of a business process. A single workflow may either be sequential, with each step contingent upon completion of the previous one, or parallel, with multiple steps occurring simultaneously. Multiple combinations of single workflows may be connected to achieve a resulting overall process.

Business Process Re-engineering

Business Process Re-engineering (BPR) was originally conceptualized by Hammer and Davenport as a means to improve organizational effectiveness and productivity. It consisted of starting from a blank slate and completely recreating major business processes as well as the use of information technology for significant performance improvement. The term unfortunately, became associated with corporate "downsizing" in the mid-1990s.

Business Process Management (BPM)

Business Process Management also termed as BPM covers how we study, identify, change and monitor business processes to ensure they run smoothly and can be improved over time. It is a continuous evaluation of existing processes and identification of ways to improve upon it, resulting in a cycle of overall organizational improvement.

Knowledge Management

Knowledge Management is the definition of the knowledge that employees and systems use to perform their functions and maintaining it in a format that can be accessed by others. The Gartner Group definition states that "Knowledge management is a discipline that promotes an integrated approach to identifying, capturing, evaluating, retrieving, and sharing all of an enterprise's information assets. These assets may include databases, documents, policies, procedures, and previously un-captured expertise and experience in individual workers."

Total Quality Management

Total Quality Management (TQM) emerged in the early 1980s as organizations sought to improve the quality of their products and services. It was followed by the Six Sigma methodology in the mid-1980s, first introduced by Motorola. Six Sigma consists of statistical methods to improve business processes and thus reduce defects in outputs. The "lean approach" to quality management was introduced by the Toyota Motor Company in the 1990s and focused on customer needs and reduction of wastage.

Information Technology as an Enabler for Business Process Management

Advances in information technology over the years, have changed business processes within and between business enterprises. In the 1960s, operating systems had limited functionality and any workflow management systems that were in use, were tailor made for the specific organization.

The 1970s-1980s saw the development of data-driven approaches, as data storage and retrieval technologies improved. Data modeling rather than process modeling was the starting point for building an information system. Business processes had to adapt to information technology because process modeling was neglected.

The shift towards process oriented management occurred in the 1990s. Enterprise resource planning software with workflow management components such as SAP, Baan, PeopleSoft, Oracle and JD Edwards emerged.

The world of e-business created a need to automate business processes across organizations, which in turn raised the need for standardized protocols and web services composition languages that can be understood across the industry. The Business Process Modeling Notation (BPMN) and Business Motivation Model (BMM) are widely used standards for business modeling. The Business Modeling and Integration Domain Task Force (BMI DTF) is a consortium of vendors and user companies that continues to work together to develop standards and specifications to promote collaboration and integration of people, systems, processes and information within and across enterprises.

The most recent trends in BPM are influenced by emergence of cloud technology, the prevalence of social media, mobile technology and development of analytical techniques. Cloud based technologies allow companies to purchase resources quickly and as required independent of their location. Social media, websites and smart phones are the newest channels through which organizations reach and support their customers. The abundance of customer data collected through these channels as well as through call center interactions, emails, voice calls, and customer surveys has led to a huge growth in data analytics which in turn is utilized for performance management and improving the ways in which the company services its customers.

Importance of the Process Chain

Business processes comprise a set of sequential sub-processes or tasks with alternative paths, depending on certain conditions as applicable, performed to achieve a given objective or produce given outputs. Each process has one or more needed inputs. The inputs and outputs may be received from, or sent to other business processes, other organizational units, or internal or external stakeholders.

Business processes are designed to be operated by one or more business functional units, and emphasize the importance of the "process chain" rather than the individual units.

In general, the various tasks of a business process can be performed in one of two ways

1. manually and
2. by means of business data processing systems such as ERP systems.

Typically, some process tasks will be manual, while some will be computer-based, and these tasks may be sequenced in many ways. In other words, the data and information that are being handled through the process may pass through manual or computer tasks in any given order.

Policies, Processes and Procedures

The above improvement areas are equally applicable to policies, processes and detailed procedures (sub-processes/tasks). There is a cascading effect of improvements made at a higher level on those made at a lower level.

For instance, if a recommendation to replace a given policy with a better one is made with proper justification and accepted in principle by business process owners, then corresponding changes in the consequent processes and procedures will follow naturally in order to enable implementation of the policies

Manual / Administrative Vs. Computer System-based Internal Controls

Internal controls can be built into manual / administrative process steps and / or computer system procedures.

It is advisable to build in as many system controls as possible, since these controls, being automatic, will always be exercised since they are built into the design of the business system software. For instance, an error message preventing an entry of a received raw material quantity exceeding the purchase order quantity by greater than the permissible tolerance percentage will always be displayed and will prevent the system user from entering such a quantity.

However, for various reasons such as practicality, the need to be "flexible" (whatever that may signify), lack of business domain knowledge and experience, difficulties in designing/writing software, cost of software development/modification, the incapability of a computerised system to provide controls, etc., all internal controls otherwise considered to be necessary are often not built into business systems and software.

In such a scenario, the manual, administrative process controls outside the computer system should be clearly documented, enforced and regularly exercised. For instance, while entering data to create a new record in a material system database's item master table, the only internal control that the system can provide over the item description field is not to allow the user to leave the description blank – in other words, configure item description as a mandatory field. The system obviously cannot alert the user that the description is wrongly spelled, inappropriate, nonsensical, etc.

In the absence of such a system-based internal control, the item creation process must include a suitable administrative control through the detailed checking, by a responsible officer, of all fields entered for the new item, by comparing a print-out taken from the system with the item data entry sheet, and ensuring that any corrections in the item description (and other similar fields where no system control is possible) are promptly carried out.

Last but not least, the introduction of effective manual, administrative controls usually requires an overriding periodic check by a higher authority to ensure that such controls are exercised in the first place.

Information Reports as an Essential Base for Execution

Business processes must include up-to-date and accurate reports to ensure effective action. An example of this is the availability of purchase order status reports for supplier delivery follow-up as described in the section on effectiveness above. There are numerous examples of this in every possible business process.

Another example from production is the process of analysis of line rejections occurring on the shop floor. This process should include systematic periodical analysis of rejections by reason, and present the results in a suitable information report that pinpoints the major reasons, and trends in these reasons, for management to take corrective actions to control rejections and keep them within acceptable limits. Such a process of analysis and summarisation of line rejection events is clearly superior to a process which merely inquires into each individual rejection as it occurs.

Business process owners and operatives should realise that process improvement often occurs with introduction of appropriate transaction, operational, highlight, exception or M.I.S. reports, provided these are consciously used for day-to-day or periodical decision-making. With this understanding would hopefully come the willingness to invest time and other resources in business process improvement by introduction of useful and relevant reporting systems.

Supporting Theories and Concepts

Span of Control

The span of control is the number of subordinates a supervisor manages within a structural organization. Introducing a business process concept has a considerable impact on the structural elements of the organization and thus also on the span of control.

Large organizations that are not organized as markets need to be organized in smaller units - departments - which can be defined according to different principles.

Information Management Concepts

Information Management and the organization design strategies being related to it, are a theoretical cornerstone of the business process concept.

Artifact-centric Business Process Model

Artifact-centric business process model represents an operational model of business processes in which the changes and evolution of business data, or business entities, are considered as the main driver of the processes. The artifact-centric approach, also called activity-centric or process-centric business process modeling, focuses on describing how business data is change/updated, by a particular action or task, throughout the process.

Overview

In general, a process model describes activities conducted in order to achieve business goals, informational structures, and organizational resources. Workflows, as a typical process modeling approach, often emphasize the sequencing of activities (i.e., control flows), but ignore the informational perspective or treat it only within the context of single activities. Without a complete view of the informational context, business actors often focus on what should be done instead of what can be done, hindering operational innovations.

Business process modeling is a foundation for design and management of business processes. Two key aspects of business process modeling are a formal framework that integrates both control flow and data, and a set of tools to assist all aspects of a business process life cycle. A typical business process life cycle includes at least a design phase, concerned with the "correct" realization of business logic in a resource-constrained environment, and an operational phase, concerned with optimizing and improving execution (operations). Traditional business process models emphasize a procedural and/or graph-based paradigm (i.e., control flow). Thus, methodologies to design workflow in those models are typically process-centric. It has been argued that a data-centric perspective is more useful for designing business processes in the modern era.

Intuitively, business artifacts (or simply artifacts) are data objects whose manipulations define the underlying processes in a business model. Recent engineering and development efforts have adopted the artifact approach for design and analysis of business models. An important distinction between artifact-centric models and traditional data flow (computational) models is that the notion of the life cycle of the data objects is prominent in the former, while not existing in the latter.

Research and History

Artifact-centric modeling is an area of growing interest. Nigam and Caswell introduced the concept of business artifacts and information-centric processing of artifact lifecycles. Kumaran et al.'s further studies on artifact-centric business processes can be found here. Bhattacharya described a successful business engagement which applies business artifact techniques to industrialize discovery processes in pharmaceutical research. Liu et al. formulated nine commonly used patterns in information-centric business operation models and developed a computational model based on Petri Nets. Bhattacharya, K., et al. provides a formal model for artifact-centric business processes with complexity results concerning static analysis of the semantics of such processes. Kumaran et al. presented the formalized information-centric approach to discovering business entities from activity-centric process models and transforming such models into artifact-centric business process models. An algorithm was provided to achieve this transformation automatically.

Other approaches related to artifact-centric modelling can be found in,. Van der Aalst et al. provides a case-handling approach where a process is driven by the presence of data objects instead of control flows. A case is similar to the business entity concept in many respects. Wang and Kumar proposed the document-driven workflow systems which is designed based on data dependencies without the need for explicit control flows. Muller et al. also introduced the framework for the data-driven modelling of large process structures, namely COREPRO. The approach reduces modelling efforts significantly and provides mechanisms for maintaining data-driven process structures.

Another related thread of work is the use of state machines to model object lifecycles. Industries often define data objects and standardize their lifecycles as state machines to facilitate interoperability between industry partners and enforce legal regulations. Redding et al. and Küster et al. give techniques to generate business processes which are compliant with predefined object lifecycles. In addition, event-driven process modelling, for example, Event-driven Process Chains (EPC), also describes object lifecycles glued by events.

More recent and closely related work on artifact-centric process model can be found in. Gerede and Su developed a specification language ABSL to specify artifact behaviours in artifact-centric process models. The authors showed decidability results of our language for different cases and provided key insights on how artifact-centric view can affect the specification of desirable business properties. Gerede et al. identified important classes of properties on artifact-centric operational models focusing on persistence, uniqueness and arrival properties. They proposed a formal model for artifact-centric operational models to enable a static analysis of these properties and showed that the formal model guarantees persistence and uniqueness.

Fritz, Hull, and Su formulated the technical problem of goal-directed workflow construction in the context of declarative artifact-centric workflow, and develop results concerning the general setting, design time analysis, and the synthesis of workflow schemas from goal specifications. The work is among the important initial steps along the path towards eventual support for tools that enable substantial automation for workflow design, analysis, and modification. Deutsch et al. introduced the artifact system model, which formalizes a business process modelling paradigm that has recently attracted the attention of both the industrial and research communities. The problem of automatic verification of artifact systems, with the goal of increasing confidence in the correctness of such business processes is also studied.

Sira and Chengfei proposed a novel view framework for artifact-centric business processes. It consists of artifact-centric process model, process view model, a set of consistency rules, and the construction approach for building process views. The formal model of artifact-centric business processes and views, namely ACP, is defined and used to describe artifacts, services, business rules that control the processes, as well as views. They developed a bottom-up abstraction mechanism for process view construction to derive views from underlying process models according to view requirements. Consistency rules are also defined to preserve the consistency between constructed view and its underlying process. This work can be considered as one approach to the abstraction, i.e., generalization of artifact-centric business processes. The framework has also been extended to address modelling and change validation of inter-organizational business processes.

Event-driven Process Chain

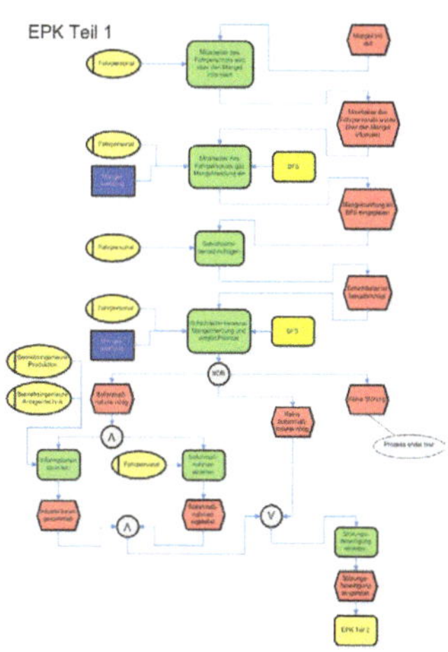

Example of a more complex EPC diagram (in German).

An Event-driven Process Chain (EPC) is a type of flowchart used for business process modelling. EPCs can be used for configuring an enterprise resource planning (ERP) implementation, and for business process improvement. Usage for control of work share with instances of autonomous workflows in workflow management is possible, but not yet implemented.

The event-driven process chain method was developed within the framework of Architecture of Integrated Information Systems (ARIS) by August-Wilhelm Scheer at the Institut für Wirtschaftsinformatik, Universität des Saarlandes (Institute for Business Information Systems at the University of Saarland) in the early 1990s.

Overview

Businesses use event-driven process chain diagrams to lay out business process workflows, originally in conjunction with SAP R/3 modeling, but now more widely. It is used by many companies for modeling, analyzing, and redesigning business processes. The Event-driven process chain method was developed within the framework of Architecture of Integrated Information Systems (ARIS). As such it forms the core technique for modeling in ARIS, which serves to link the different views in the so-called control view. To quote from a 2006 publication on Event-driven Process Chains:

> *An Event-driven process chain (EPC) is an ordered graph of events and functions. It provides various connectors that allow alternative and parallel execution of processes. Furthermore it is specified by the usages of logical operators, such as OR, AND, and XOR. A major strength of EPC is claimed to be its simplicity and easy-to-understand notation. This makes EPC a widely acceptable technique to denote business processes.*

The statement that event-driven process chains are ordered graphs is also found in other directed graphs for which no explicit node ordering is provided. No restrictions actually appear to exist on the possible structure of EPCs, but nontrivial structures involving parallelism have ill-defined execution semantics; in this respect they resemble UML activity diagrams.

Several scientific articles are devoted to providing well-defined execution semantics for general Event-driven Process Chains. One particular issue is that EPCs require non-local semantics, i.e., the execution behavior of a particular node within an EPC may depend on the state of other parts of the EPC, arbitrarily far away.

Elements of an Event-driven Process Chain

In the following the elements used in event-driven process chain diagram will be described:

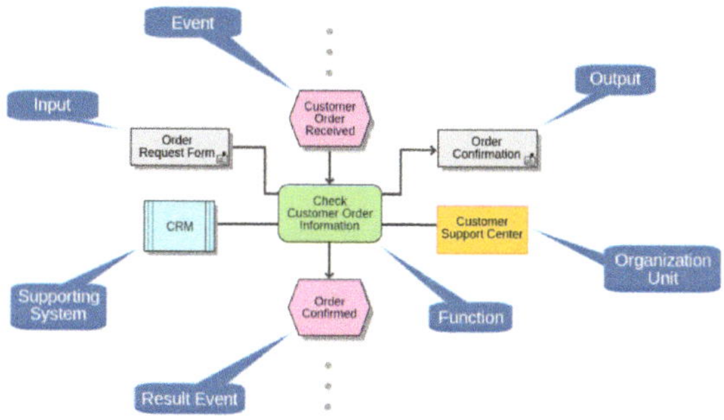

Elements of an Event-driven Process Chain

Event

Events are passive elements in event-driven process chains. They describe under what circumstances a function or a process works or which state a function or a process results in. Examples of events are "requirement captured", "material in stock", etc. In the EPC graph an event is represented as hexagon. In general, an EPC diagram must start with an event and end with an event.

Function

Functions are active elements in an EPC. They model the tasks or activities within the company. Functions describe transformations from an initial state to a resulting state. If different resulting states can occur, the selection of the respective resulting state can be modeled explicitly as a decision function using logical connectors. Functions can be refined into another EPC. In this case it is called a hierarchical function. Examples of functions are "capture requirement", "check material in stock", etc. In the event-driven process chain graph a function is represented as rounded rectangle.

Process owner

Process owner is responsible for a function (i.e. a booking clerk is responsible for booking journeys). The process owner is usually part of an organization unit (i.e. a booking clerk belongs to the booking department). It is represented as a square with a vertical line.

Organization unit

Organization units determine which organization within the structure of an enterprise is responsible for a specific function. Examples are "sales department", "procurement department", etc. It is represented as an ellipse with a vertical line.

Information, material, or resource object

In the event-driven process chain, the information, material, or resource objects portray objects in the real world, for example business objects, entities, etc., which can be input data serving as the basis for a function, or output data produced by a function. Examples are "material", "order", etc. In the EPC graph such an object is represented as rectangle.

Logical connector

In the event-driven process chain the logical relationships between elements in the control flow, that is, events and functions are described by logical connectors. With the help of logical connectors it is possible to split the control flow from one flow to two or more flows and to synchronize the control flow from two or more flows to one flow.

Logical relationships

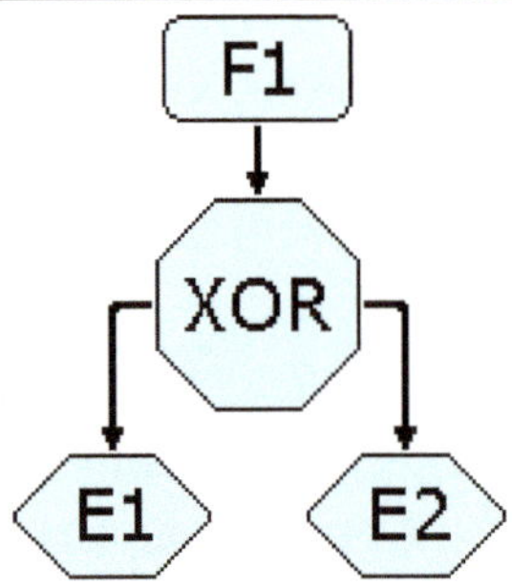	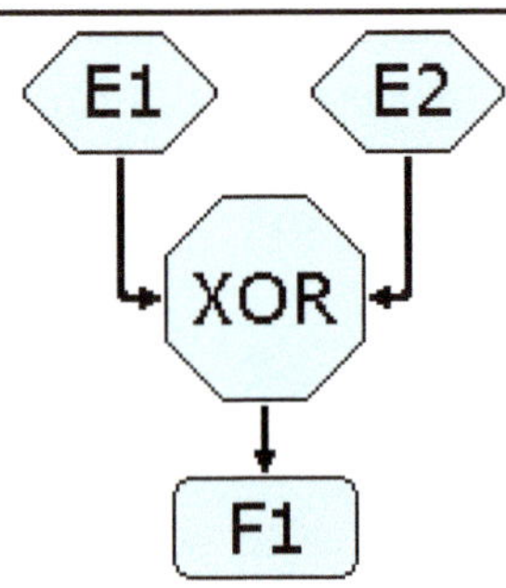
If function F1 completes, either events E1 or E2 occur	If either events E1 or E2 occur, function F1 starts

There are three kinds of logical relationships defined in event-driven process chains:

- Branch/Merge: Branch and merge correspond to making decision of which path to choose among several control flows. A branch may have one incoming control flow and two or more outgoing control flows. When the condition is fulfilled, a branch activates exactly only one of the outgoing control flows and deactivates the others. The counterpart of a branch is a merge. A merge may have two or more incoming flows and one outgoing control flow. A merge synchronizes an activated and the deactivated alternatives. The control will then be passed to the next element after the merge. A branch in the EPC is represented by an opening XOR, whereas a merge is represented as a closing XOR connectors.
- Fork/Join : Fork and join correspond to activating all paths in the control flow concurrently. A fork may have one incoming control flow and two or more outgoing control flows. When the condition is fulfilled, a fork activates all of the outgoing control flows in parallel. A join may have two or more incoming control flows and one outgoing control flow. A join synchronizes all activated incoming control

flows. In the Event-driven Process Chain diagram how the concurrency achieved is not a matter. In reality the concurrency can be achieved by true parallelism or by virtual concurrency achieved by interleaving. A fork in the EPC is represented by an opening 'AND', whereas a join is represented as a closing 'AND' connectors.

- OR : An 'OR' relationship corresponds to activating one or more paths among control flows. An opening 'OR' connector may have one incoming control flow and two or more outgoing control flows. When the condition is fulfilled, an opening 'OR' connector activates one or more control flows and deactivates the rest of them. The counterpart of this is the closing 'OR' connector. When at least one of the incoming control flows is activated, the closing 'OR' connector will pass the control to the next element after it.

Control flow

A control flow connects events with functions, process paths, or logical connectors creating chronological sequence and logical interdependencies between them. A control flow is represented as a dashed arrow.

Information flow

Information flows show the connection between functions and input or output data, upon which the function reads changes or writes.

Organization unit assignment

Organization unit assignments show the connection between an organization unit and the function it is responsible for.

Process path

Process paths serve as navigation aid in the EPC. They show the connection from or to other processes. The process path is represented as a compound symbol composed of a function symbol superimposed upon an event symbol. To employ the process path symbol in an Event-driven Process Chain diagram, a symbol is connected to the process path symbol, indicating that the process diagrammed incorporates the entirety of a second process which, for diagrammatic simplicity, is represented by a single symbol.

Example

As shown in the example, a customer order received is the initial event which creates a requirement capture within the company. In order to specify this function, sales is responsible for marketing, currency etc. As a result, event 'requirement captured' leads to another new function: check material on stock, in order to manufacture the productions.

All input or output data about material remains in the information resource. After checking material, two events may happen-with or without material on stock. If positive, get material from stock; if not, order material from suppliers. Since the two situations cannot happen at the same time, XOR is the proper connector to link them together.

Tools

There are a number of tools for creating EPC diagrams:

- BIC Platform by GBTEC
- ARIS Platform by Software AG,
- Free (Registration necessary) modeling tool ARIS Express by Software AG,
- Open Source modeling tool bflow (Java necessary),
- ADONIS of BOC Group
- Mavim Rules of Mavim BV,
- Visual Paradigm of Visual Paradigm Int.,
- Visio of Microsoft Corp.,
- Semtalk of Semtation GmbH, or
- Bonapart by Pikos GmbH.
- ConceptDraw PRO with EPC Solution

Some but not all of these tools support the tool-independent Event-driven Process Chain Markup Language (EPML) interchange format. There are also tools that generate EPC diagrams from operational data, such as SAP logs. EPC diagrams use symbols of several kinds to show the control flow structure (sequence of decisions, functions, events, and other elements) of a business process.

Meta-model of Event-driven Process Chain

Although a real process may include a series of stages until it is finished eventually, the main activities remain similar. An event triggers one function; and a function will lead to one event. Meanwhile, an event may involve one or more processes to fulfill but a process is unique for one event, the same goes for process and process path.

As for the function, its data may be included in one or more information resources, while organization unit is only responsible for one specific function.

Disintermediation

In economics, disintermediation is the removal of intermediaries from a supply chain, or "cutting out the middlemen" in connection with a transaction or a series of transactions. Instead of going through traditional distribution channels, which had some type of intermediary (such as a distributor, wholesaler, broker, or agent), companies may now deal with customers directly, for example via the Internet.

Disintermediation may decrease the total cost of servicing customers and may allow the manufacturer to increase profit margins and/or reduce prices.

Disintermediation initiated by consumers is often the result of high market transparency, in that buyers are aware of supply prices direct from the manufacturer. Buyers may choose to bypass the middlemen (wholesalers and retailers) to buy directly from the manufacturer, and pay less. Buyers can alternatively elect to purchase from wholesalers. Often, a business-to-consumer electronic commerce (B2C) company functions as the bridge between buyer and manufacturer.

However manufacturers will still incur distribution costs, such as the physical transport of goods, packaging in small units, advertising, and customer helplines, some or all of which would previously have been borne by the intermediary.

To illustrate, a typical B2C supply chain is composed of four or five entities (in order):

- Supplier
- Manufacturer
- Wholesaler
- Retailer
- Buyer

It has been argued that the Internet modifies the supply chain due to market transparency:

- Supplier
- Manufacturer
- Buyer

History

The term was originally applied to the banking industry in 1967; disintermediation occurred when consumers avoided the intermediation of banks by investing directly in securities (government and private bonds, and stocks) rather than leaving their money in savings accounts. The original cause was a U.S. government regulation (Regulation Q) which limited the interest rate paid on interest bearing accounts that were insured by the Federal Deposit Insurance Corporation.

It was later applied more generally to "cutting out the middleman" in commerce, though the financial meaning remained predominant. Only in the late 1990s did it become widely popularized.

Impact of Internet-related Disintermediation Upon Various Industries

Disintermediation has acquired a new meaning with the advent of the virtual marketplace. The virtual marketplace sellers like Amazon are edging out the middlemen. It must be understood, however, that direct sellers and buyers connect with each other because of the platform created by the virtual marketplace vendor. There is quid pro quo for the vendor for the use of the platform, else it would make no business sense to create such a platform. If the buyer, having connected with the seller, circumvents the platform and talks to the buyer and does her deal directly with the buyer, then the platform owner is unlikely to get her revenue share. This may be considered a new form of disintermediation.

Strong Impact

- Computer hardware and software
- Travel agencies
- Bookstores and music stores
- Stock Purchasing
- Contact lenses

Still in Progress (Due to Legal or Structural Obstacles)

- Alcoholic beverages
- Real estate
- Recruitment
- Education
- Banking - due to developments in cryptocurrencies and specifically Bitcoin
- Taxicab driver - due to developments in mobile applications and specifically Uber

- Hotels - due to developments in mobile applications and specifically Air BnB
- Music Industry - due to developments in mp3 and specifically BitTorrent protocol encryption

A real estate market is social. Buyers and sellers communicate to discover information and negotiate to exchange goods or services. Internet transparency is letting home buyers view Residential and Commercial MLS, FSBO listings on their own. It has reduced home buyers search costs, and given them access to multiple new product options to choose from when entering a real estate transaction. Sellers have found new tools and services to attract home buyers and sell their houses, they can now leverage Internet market tools that are intended to increase the efficacy of transactional requirements.

This transparency has made it difficult for real estate agents, appraisers, and lenders to collect the fees, tipping the balance of power towards the consumers. By opening access to information outside of the Brokers/Lawyers control, buyers and sellers now gain economic benefits that would otherwise go to market intermediaries or be inappropriately distributed among the smart and connected dealmakers of the financial world.

- Prime brokerage for hedge funds
- Health Care - due to developments in eHealth and specifically Consumer Health Informatics
- Video rental and distribution

Failed and Became Niche Ancillary Services

- Furniture
- Groceries
- Pet supplies (especially dog food)

Discussion

In the non-Internet world, disintermediation has been an important strategy for many big box retailers like Walmart, which attempt to reduce prices by reducing the number of intermediaries between the supplier and the buyer. Disintermediation is also closely associated with the idea of just in time manufacturing, as the removal of the need for inventory removes one function of an intermediary.

The existence of laws which discourage disintermediation has been cited as a reason for the poor economic performance of Japan and Germany in the 1990s.

However, Internet-related disintermediation occurred less frequently than many expected during the dot com boom. Retailers and wholesalers provide essential functions such as the extension of credit, aggregation of products from different suppliers, and processing of returns. In addition, shipping goods to and from the manufacturer can in many cases be far less efficient than shipping them to a store where the consumer can pick them up (if the consumer's trip to the store is ignored). In response to the threat of disintermediation, some retailers have attempted to integrate a virtual presence and a physical presence in a strategy known as bricks and clicks.

Reintermediation

Reintermediation can be defined as the reintroduction of an intermediary between end users (consumers) and a producer. This term applies especially to instances in which disintermediation has occurred first.

At the start of the Internet revolution, electronic commerce was seen as a tool of disintermediation for cutting operating costs. The concept was that by allowing consumers to purchase products directly from producers via the Internet, the product delivery chain would be drastically shortened, thereby "disintermediating" the standard supply model middlemen. However, what largely happened was that new intermediaries appeared in the digital landscape (e.g., Amazon.com and eBay).

Reintermediation occurred due to many new problems associated with the e-commerce disintermediation concept, largely centered on the issues associated with the direct-to-consumers model. The high cost of shipping many small orders, massive customer service issues, and confronting the wrath of disintermediated retailers and supply channel partners all presented real obstacles. Huge resources are required to accommodate presales and postsales issues of individual consumers. Before disintermediation, supply chain middlemen acted as salespeople for the producers. Without them, the producer itself would have to handle procuring those customers. Selling online has its own associated costs: developing quality websites, maintaining product information, and marketing expenses all add up. Finally, limiting a product's availability to Internet channels forces the producer to compete with the rest of the Internet for customers' attention, a space that is becoming increasingly crowded.

Examples of Companies

Notable examples of disintermediation include Dell and Apple, which sell many of their systems direct to the consumer—thus bypassing traditional retail chains, having succeeded in creating brands well recognized by customers, profitable and with continuous growth.

Direct Selling

Direct selling is the marketing and selling of products directly to consumers away from a fixed retail location. Peddling is the oldest form of direct selling. Modern direct selling includes sales made through the party plan, one-on-one demonstrations, and other personal contact arrangements as well as internet sales. A textbook definition is: "The direct personal presentation, demonstration, and sale of products and services to consumers, usually in their homes or at their jobs."

According to the WFDSA, consumers benefit from direct selling because of the convenience and service it provides, including personal demonstration and explanation of products, home delivery, and generous satisfaction guarantees. In contrast to franchising, the cost for an individual to start an independent direct selling business is typically very low with little or no required inventory or other cash commitments to begin.

Most direct selling associations around the world require their members to abide by a code of conduct towards a fair partnership both with customers and salesmen...

Most national direct selling associations are represented in the World Federation of Direct Selling Associations (WFDSA).

Direct selling is distinct from direct marketing because it is about individual sales agents reaching and dealing directly with clients. Direct marketing is about business organizations seeking a relationship with their customers *without* going through an agent/consultant or retail outlet.

Direct selling consists of two main business models: single-level marketing, in which a direct seller makes money by buying products from a parent organization and selling them directly to customers, and multi-level marketing (also known as *network marketing* or *person-to-person marketing*), in which the direct seller makes money from both direct sales to customers and by sponsoring new direct sellers and earning a commission from their efforts.

Largest Direct Selling Companies

According to Direct Selling News, the largest direct selling companies globally, by revenue in 2016, were:

2016 Rank	Company Name	2015 Revenue
1	Amway	US$ 9.50 B
2	Avon Products	US$ 6.16 B
3	Herbalife	US$ 4.47 B
4	Vorwerk	US$ 4.00 B
5	Infinitus	US$ 3.88 B
6	Mary Kay	US$ 3.70 B

7	Perfect	US$ 3.58 B
8	Natura	US$ 2.41 B
9	Tupperware	US$ 2.28 B
10	Nu Skin	US$ 2.25 B

Freemium

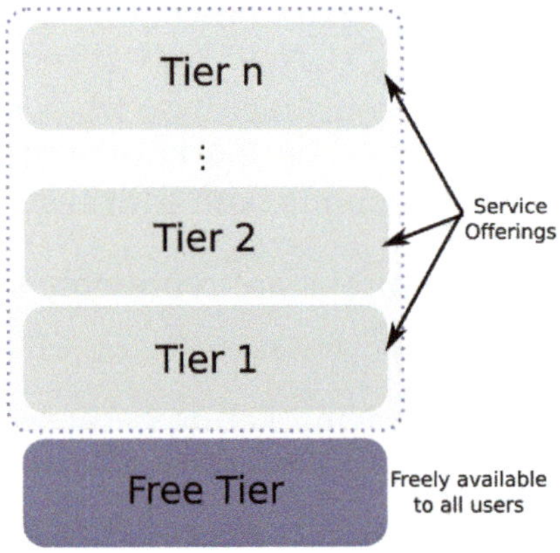

In freemium business model, business tiers start with a "free" tier

Freemium is a pricing strategy by which a product or service (typically a digital offering or application such as software, media, games or web services) is provided free of charge, but money (premium) is charged for proprietary features, functionality, or virtual goods.

Origin

The business model called shareware has been in use for software (crippleware or freeware) since the 1980s, particularly in the form of a free time- or feature-limited ("lite") version, often given away on a floppy disk or CD-ROM, to promote a paid-for full version. The model is particularly suited to software as the cost of distribution is negligible, so little is lost by giving it away for free – as long as significant cannibalization is avoided.

The term *freemium* to describe this model appears to have been created only much later, in response to a 2006 blog post by venture capitalist Fred Wilson summarizing the model:

Give your service away for free, possibly ad supported but maybe not, acquire a lot of customers very efficiently through word of mouth, referral networks, organic search marketing, etc., then offer premium priced value added services or an enhanced version of your service to your customer base.

Jarid Lukin of Alacra, one of Wilson's portfolio companies, then suggested the term "freemium" for this model.

In 2009, Chris Anderson published the book *Free*, which examines the popularity of this business model. As well as for traditional software and services, it is now also often used by Web 2.0 and open source companies. In 2014, Eric Seufert published the book *Freemium Economics*, which attempts to deconstruct the economic principles of the freemium model and prescribe a framework for implementing them into software products.

The freemium model is closely related to tiered services. It has become a highly popular model, with notable examples including LinkedIn, Badoo, and in the form of a "soft" paywall, such as those employed by *The New York Times* and by Press+. A freemium model is sometimes used to build a consumer base when the marginal cost of producing extra units is low.

In the Christian 503 (c) 3 religious education sector, Christian Leaders Institute (CLI) uses the freemium model to train Christian leaders. CLI offers free training to receive "unofficial" certificates, diplomas and degrees at no charge. The infrastructure cost to provide this training and these "unofficial" awards is supported by donations to the mission, which receive tax deductions in the United States. If CLI students want official awards, they are required to make minimal "non-tax deductible" donations that cover the costs of individual processing, verification, and delivery.

Cloud service providers – ranging from Infrastructure providers (IaaS), platform providers (PaaS) to software service providers (SaaS) – use freemium pattern in their business models. Companies such as Rackspace, Heroku, and Amazon use freemium business patterns by providing cloud services for free, but in limited and trial use. Amazon, for example, allows anyone to create a simple cloud in their infrastructure as a trial service for 12 months.

Other examples include free-to-play games – video games that can be downloaded without paying. Video game publishers of free-to-play games rely on other means to generate revenue – such as optional in-game items that can be purchased by players to enhance game-play and/or aesthetics.

Restrictions

Ways in which the product or service may be restricted in the free version include:

- Feature limited (e.g. a "lite" version of software, such as not including features like three-way video calling)
- Capacity limited (e.g. SQL Server Express, which is restricted to databases of 10GB or less)
- Seat limited (e.g. only usable on one computer rather than across a network)

- Customer class limited (e.g. only usable by educational users; most Autodesk software with full features is free for students)
- Effort limited (e.g. Temple Run, in which all or most features are available for free, but require extended unlocking or slowly obtained in-game currency which can be accelerated or purchased for a fee)
- Support limited (e.g. users of a "lite" version do not receive telephone and/or email support)
- Time or bandwidth limited
- Storage Space Limited (e.g. free initial Storage of a Cloud storage provider, additional storage on a subscription model thereafter)

Some software and services make all of the features available for free for a trial period, and then at the end of that period revert to operating as a feature limited free version (e.g. Online Armor Personal Firewall).The user can unlock the premium features on payment of a licence fee, as per the freemium model. Some businesses use a variation of the model known as "open core", in which the unsupported, feature-limited free version is also open source software, but versions with additional features and official support are commercial software.

Other software makes its features available for trial period, and then at the end of that period it stops working (e.g. Microsoft Office 30-day trial). This is a time-limited evaluation, and at the end of the evaluation period does not work at all. Where the user has access to a limited free version without time restraint.

Significance

In June 2011, *PC World* reported that traditional anti-virus software had started to lose market share to freemium anti-virus products. By September 2012, all but two of the 50 highest-grossing apps in the Games section of Apple's iTunes App Store supported in-app purchases, leading *Wired* to conclude that game developers were now required to choose between including such purchases or foregoing a very substantial revenue stream. Beginning in 2013, the digital distribution platform Steam began to add numerous Free to play and Early Access games to its library, many of which utilized Freemium marketing for their in-game economies. Due to criticism that the multiplayer games falling under this were "pay to win" in nature or were low-quality and never finished development, Valve Corporation has since added stricter rules to its Early Access and Free-to-play policies.

Criticism of Freemium Games

Freemium games have come under criticism from players and critics. Many are pinned with the derogatory term 'Pay2Win', a term which criticizes freemium games for giving an advantage to players who pay more money as opposed to those who have more skill.

Criticisms also extend to the way that the business model can often appear unregulated, to the point of encouraging prolific spending. An example of this is *Smurfs' Village*, which was released by Capcom as a free game tied in with the release of the 2011 film. The game was very popular, at one point overtaking *Angry Birds* as the top-downloaded app on the Apple iOS App Store. The game had the player buy characters and buildings using in-game currency, but more currency could be purchased with real money using account details tied to the device, using Apple's store authentication prior to purchase. Prior to changes Apple made in 2011, it was possible to use the authentication for up to fifteen minutes before having to reconfirm the password, and parents had soon found that their children had used this to unwittingly incur hundreds of US dollars in charges by playing the game. A group of parents filed suit against Apple to change their practice in microtransactions to prevent this from happening in other freemium games.

In November, 2014, the sixth episode of Season 18 of the animated TV series *South Park* aired an episode entitled "Freemium Isn't Free." The episode satirized the business model for encouraging predatory game design tactics based on crooked business model.

In 2015, Nintendo released two of their own freemium games in the *Pokémon* series based on other standalone purchasable titles. Nintendo has taken a different approach with freemium games by making it possible to complete the entire game without buying premium credits, having that as an option so players can proceed through the game at a pace that suits them.

References

- Wim F. Gielingh, Alexander K. Suhm, Michael Böhms (1993). IMPPACT Reference Model. Springer ISBN 3540561501 p.37.
- Daniel A. Menascé, Virgilio A. F. Almeida (2000). Scaling for E-business: Technologies, Models, Performance, and Capacity Planning. Prentice Hall. ISBN 0130863289
- Popp, Dr. Karl Michael; Meyer, Ralf (2010). Profit from Software Ecosystems: Business Models, Ecosystems and Partnerships in the Software Industry. Norderstedt, Germany: Books on Demand. ISBN 9783839169834.
- Popp, Dr. Karl Michael (2015). Best Practices for commercial use of open source software. Norderstedt, Germany: Books on Demand. ISBN 978-3738619096.
- Solatan, Jean (2011). Advances in software economics: A reader on business models and Partner Ecosystems in the software industry. Norderstedt, Germany: BOD. ISBN 978-3-8448-0405-8.
- Xardel, Dominique (1993). The Direct Selling Revolution. Understanding the Growth of the Amway Corporation. Blackwell Publishing. pp. 3–4. ISBN 978-0-631-19229-9.
- Bridge, Rachel; Edwards, Paul; Edwards, Sarah A.; Economy, Peter (2014). Starting a Home Business for Dummies. Wiley. p. 36. ISBN 978-1-118-73757-6. Retrieved 31 December 2015.
- "Bountysource Raises $1.1 Million for the First Crowdfunding Platform for Open-Source Soft-

ware Projects". Yahoo Finance. Marketwired. 16 July 2013. Retrieved 18 June 2016.

- Wheeler, David A. (February 2009). "F/LOSS is Commercial Software". Technology Innovation Management Review. Talent First Network. Retrieved 18 June 2016.
- Germain, Jack M. (5 November 2013). "FOSS in the Enterprise: To Pay or Not to Pay?". LinuxInsider. ECT News Network, Inc. Retrieved 18 June 2016.
- Gartner, Samantha (6 October 2014). "Moodle will always be an open source project". opensource.com. Retrieved 18 June 2016.
- Dougiamas, Martin (22 January 2014). "Moodle: a case study in sustainability". OSS Watch. University of Oxford (published 5 June 2007). Retrieved 18 June 2016.
- Tiemann, Michael (7 June 2007). "GNU Affero GPL version 3 and the "ASP loophole"". Open Source Initiative. Retrieved 18 June 2016.
- "Frequently Asked Questions about the GNU Licenses". GNU Project. Free Software Foundation. 26 May 2016. Why did you decide to write the GNU Affero GPLv3 as a separate license?. Retrieved 18 June 2016.

4

Diverse Aspects of Business Process Management

The aspects of business process management explained in the following section are business process reengineering, outsourcing, business activity monitoring, Internet of things, business process automation and virtual business model. Business process reengineering is a management strategy that focuses and analysis workflows and the process within any organization. The major components of business process management are discussed in this chapter.

Business Process Reengineering

Business process re-engineering (BPR) is a business management strategy, originally pioneered in the early 1990s, focusing on the analysis and design of workflows and business processes within an organization. BPR aimed to help organizations fundamentally rethink how they do their work in order to dramatically improve customer service, cut operational costs, and become world-class competitors. In the mid-1990s, as many as 60% of the Fortune 500 companies claimed to either have initiated reengineering efforts, or to have plans to do so.

BPR seeks to help companies radically restructure their organizations by focusing on the ground-up design of their business processes. According to Davenport (1990) a business process is a set of logically related tasks performed to achieve a defined business outcome. Re-engineering emphasized a holistic focus on business objectives and how processes related to them, encouraging full-scale recreation of processes rather than iterative optimization of sub-processes.

Business process reengineering is also known as business process redesign, business transformation, or business process change management.

Overview

Business process reengineering (BPR) is the practice of rethinking and redesigning the way work is done to better support an organization's mission and reduce costs. Reengineering starts with a high-level assessment of the organization's mission, strategic goals, and customer needs. Basic questions are asked, such as "Does our mission need to be redefined? Are our strategic goals aligned with our mission? Who are our customers?" An organization may find that it is operating on questionable assumptions, particularly in terms of the wants and needs of its customers. Only after the organization rethinks what it should be doing, does it go on to decide how best to do it.

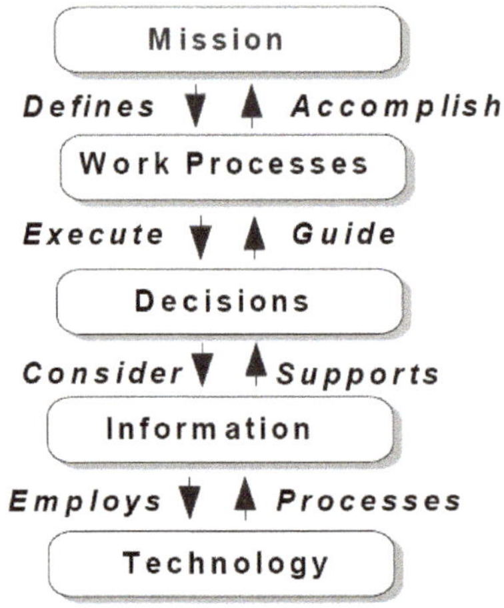

Reengineering guidance and relationship of mission and work processes to information technology.

Within the framework of this basic assessment of mission and goals, re-engineering focuses on the organization's business processes—the steps and procedures that govern how resources are used to create products and services that meet the needs of particular customers or markets. As a structured ordering of work steps across time and place, a business process can be decomposed into specific activities, measured, modeled, and improved. It can also be completely redesigned or eliminated altogether. Re-engineering identifies, analyzes, and re-designs an organization's core business processes with the aim of achieving dramatic improvements in critical performance measures, such as cost, quality, service, and speed.

Re-engineering recognizes that an organization's business processes are usually fragmented into sub-processes and tasks that are carried out by several specialized functional areas within the organization. Often, no one is responsible for the overall performance of the entire process. Reengineering maintains that optimizing the performance of sub-processes can result in some benefits, but cannot yield dramatic improvements if the process itself is fundamentally inefficient and outmoded. For that reason, re-engineering focuses on re-designing the process as a whole in order to achieve the greatest possible benefits to the organization and their customers. This drive for realizing dramatic improvements by fundamentally re-thinking how the organization's work should be done distinguishes the re-engineering from process improvement efforts that focus on functional or incremental improvement.

History

Business Process Reengineering (BPR) began as a private sector technique to help organizations fundamentally rethink how they do their work in order to dramatically improve customer service, cut operational costs, and become world-class competitors. A key stimulus for re-engineering has been the continuing development and deployment of sophisticated information systems and networks. Leading organizations are becoming bolder in using this technology to support innovative business processes, rather than refining current ways of doing work.

Reengineering Work: Don't Automate, Obliterate, 1990

In 1990, Michael Hammer, a former professor of computer science at the Massachusetts Institute of Technology (MIT), published the article "Reengineering Work: Don't Automate, Obliterate" in the *Harvard Business Review*, in which he claimed that the major challenge for managers is to obliterate forms of work that do not add value, rather than using technology for automating it. This statement implicitly accused managers of having focused on the wrong issues, namely that technology in general, and more specifically information technology, has been used primarily for automating existing processes rather than using it as an enabler for making non-value adding work obsolete.

Hammer's claim was simple: Most of the work being done does not add any value for customers, and this work should be removed, not accelerated through automation. Instead, companies should reconsider their inability to satisfy customer needs, and their insufficient cost structure. Even well established management thinkers, such as Peter Drucker and Tom Peters, were accepting and advocating BPR as a new tool for (re-) achieving success in a dynamic world. During the following years, a fast-growing number of publications, books as well as journal articles, were dedicated to BPR, and many consulting firms embarked on this trend and developed BPR methods. However, the critics were fast to claim that BPR was a way to dehumanize the work place, increase managerial control, and to justify downsizing, i.e. major reductions of the work force, and a rebirth of Taylorism under a different label.

Despite this critique, reengineering was adopted at an accelerating pace and by 1993, as many as 60% of the Fortune 500 companies claimed to either have initiated reengineering efforts, or to have plans to do so. This trend was fueled by the fast adoption of BPR by the consulting industry, but also by the study *Made in America*, conducted by MIT, that showed how companies in many US industries had lagged behind their foreign counterparts in terms of competitiveness, time-to-market and productivity.

Development After 1995

With the publication of critiques in 1995 and 1996 by some of the early BPR proponents, coupled with abuses and misuses of the concept by others, the reengineering fervor in the U.S. began to wane. Since then, considering business processes as a starting point for business analysis and redesign has become a widely accepted approach and is a standard part of the change methodology portfolio, but is typically performed in a less radical way than originally proposed.

More recently, the concept of Business Process Management (BPM) has gained major attention in the corporate world and can be considered as a successor to the BPR wave of the 1990s, as it is evenly driven by a striving for process efficiency supported by information technology. Equivalently to the critique brought forward against BPR, BPM is now accused of focusing on technology and disregarding the people aspects of change.

Topics

The most notable definitions of reengineering are:

- "... the fundamental rethinking and radical redesign of business processes to achieve dramatic improvements in critical contemporary modern measures of performance, such as cost, quality, service, and speed."
- "encompasses the envisioning of new work strategies, the actual process design activity, and the implementation of the change in all its complex technological, human, and organizational dimensions."

BPR is different from other approaches to organization development (OD), especially the continuous improvement or TQM movement, by virtue of its aim for fundamental and radical change rather than iterative improvement. In order to achieve the major improvements BPR is seeking for, the change of structural organizational variables, and other ways of managing and performing work is often considered as being insufficient. For being able to reap the achievable benefits fully, the use of information technology (IT) is conceived as a major contributing factor. While IT traditionally has been used for supporting the existing business functions, i.e. it was used for increasing organizational efficiency, it now plays a role as enabler of new organizational forms, and patterns of collaboration within and between organizations.

BPR derives its existence from different disciplines, and four major areas can be identified as being subjected to change in BPR – organization, technology, strategy, and people – where a process view is used as common framework for considering these dimensions.

Business strategy is the primary driver of BPR initiatives and the other dimensions are governed by strategy's encompassing role. The organization dimension reflects the structural elements of the company, such as hierarchical levels, the composition of organizational units, and the distribution of work between them. Technology is concerned with the use of computer systems and other forms of communication technology in the business. In BPR, information technology is generally considered as playing a role as enabler of new forms of organizing and collaborating, rather than supporting existing business functions. The people / human resources dimension deals with aspects such as education, training, motivation and reward systems. The concept of business processes – interrelated activities aiming at creating a value added output to a customer – is the basic underlying idea of BPR. These processes are characterized by a number of attributes: Process ownership, customer focus, value adding, and cross-functionality.

The Role of Information Technology

Information technology (IT) has historically played an important role in the reengineering concept. It is considered by some as a major enabler for new forms of working and collaborating within an organization and across organizational borders.

BPR literature identified several so called *disruptive technologies* that were supposed to challenge traditional wisdom about how work should be performed.

- Shared databases, making information available at many places
- Expert systems, allowing generalists to perform specialist tasks
- Telecommunication networks, allowing organizations to be centralized and decentralized at the same time
- Decision-support tools, allowing decision-making to be a part of everybody's job
- Wireless data communication and portable computers, allowing field personnel to work office independent
- Interactive videodisk, to get in immediate contact with potential buyers
- Automatic identification and tracking, allowing things to tell where they are, instead of requiring to be found
- High performance computing, allowing on-the-fly planning and revisioning

In the mid-1990s, especially workflow management systems were considered as a significant contributor to improved process efficiency. Also, ERP (enterprise resource planning) vendors, such as SAP, JD Edwards, Oracle, PeopleSoft, positioned their solutions as vehicles for business process redesign and improvement.

Research and Methodology

Although the labels and steps differ slightly, the early methodologies that were rooted in IT-centric BPR solutions share many of the same basic principles and elements. The following outline is one such model, based on the PRLC (Process Reengineering Life Cycle) approach developed by Guha. Simplified schematic outline of using a business process approach, exemplified for pharmaceutical R&D

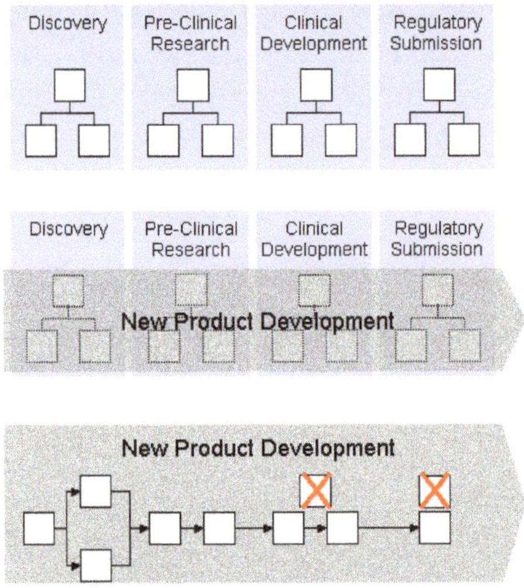

Model based on PRLC approach

1. Structural organization with functional units
2. Introduction of New Product Development as cross-functional process
3. Re-structuring and streamlining activities, removal of non-value adding tasks

Benefiting from lessons learned from the early adopters, some BPR practitioners advocated a change in emphasis to a customer-centric, as opposed to an IT-centric, methodology. One such methodology, that also incorporated a Risk and Impact Assessment to account for the effect that BPR can have on jobs and operations, was described by Lon Roberts (1994). Roberts also stressed the use of change management tools to proactively address resistance to change—a factor linked to the demise of many reengineering initiatives that looked good on the drawing board.

Some items to use on a process analysis checklist are: Reduce handoffs, Centralize data, Reduce delays, Free resources faster, Combine similar activities. Also within the management consulting industry, a significant number of methodological approaches have been developed.

Framework

An easy to follow seven step INSPIRE framework is developed by Bhudeb Chakravarti which can be followed by any Process Analyst to perform BPR. The seven steps of the framework are Initiate a new process reengineering project and prepare a business case for the same; Negotiate with senior management to get approval to start the process reengineering project; Select the key processes that need to be reengineered; Plan the process reengineering activities; Investigate the processes to analyze the problem areas; Redesign the selected processes to improve the performance and Ensure the successful implementation of redesigned processes through proper monitoring and evaluation.

Factors for Success and Failure

Factors that are important to BPR success include:

1. BPR team composition.
2. Business needs analysis.
3. Adequate IT infrastructure.
4. Effective change management.
5. Ongoing continuous improvement

The aspects of a BPM effort that are modified include organizational structures, management systems, employee responsibilities and performance measurements, incentive systems, skills development, and the use of IT. BPR can potentially affect every aspect of how business is conducted today. Wholesale changes can cause results ranging from enviable success to complete failure.

If successful, a BPM initiative can result in improved quality, customer service, and competitiveness, as well as reductions in cost or cycle time. However, 50-70% of reengineering projects are either failures or do not achieve significant benefit.

There are many reasons for sub-optimal business processes which include:

1. One department may be optimized at the expense of another
2. Lack of time to focus on improving business process
3. Lack of recognition of the extent of the problem
4. Lack of training
5. People involved use the best tool they have at their disposal which is usually Excel to fix problems
6. Inadequate infrastructure
7. Overly bureaucratic processes
8. Lack of motivation

Many unsuccessful BPR attempts may have been due to the confusion surrounding BPR, and how it should be performed. Organizations were well aware that changes needed to be made, but did not know which areas to change or how to change them. As a result, process reengineering is a management concept that has been formed by trial and error or, in other words, practical experience. As more and more businesses reengineer their processes, knowledge of what caused the successes or failures is becoming apparent. To reap lasting benefits, companies must be willing to examine how strategy and reengineering complement each other by learning to quantify strategy in terms of cost, milestones, and timetables, by accepting ownership of the strategy throughout the organization, by assessing the organization's current capabilities and process realistically, and by linking strategy to the budgeting process. Otherwise, BPR is only a short-term efficiency exercise.

Organization-wide Commitment

Major changes to business processes have a direct effect on processes, technology, job roles, and workplace culture. Significant changes to even one of those areas require resources, money, and leadership. Changing them simultaneously is an extraordinary task. Like any large and complex undertaking, implementing reengineering requires the talents and energies of a broad spectrum of experts. Since BPR can involve multiple areas within the organization, it is important to get support from all affected departments. Through the involvement of selected department members, the organization can gain valuable input before a process is implemented; a step which promotes both the cooperation and the vital acceptance of the reengineered process by all segments of the organization.

Getting enterprise wide commitment involves the following: top management sponsorship, bottom-up buy-in from process users, dedicated BPR team, and budget allocation for the total solution with measures to demonstrate value. Before any BPR project can be implemented successfully, there must be a commitment to the project by the management of the organization, and strong leadership must be provided. Reengineering efforts can by no means be exercised without a company-wide commitment to the goals. However, top management commitment is imperative for success. Top management must recognize the need for change, develop a complete understanding of what BPR is, and plan how to achieve it.

Leadership has to be effective, strong, visible, and creative in thinking and understanding in order to provide a clear vision. Convincing every affected group within the organization of the need for BPR is a key step in successfully implementing a process. By informing all affected groups at every stage, and emphasizing the positive end results of the reengineering process, it is possible to minimize resistance to change and increase the odds for success. The ultimate success of BPR depends on the strong, consistent, and continuous involvement of all departmental levels within the organization. It also depends on the people who do it and how well they can be motivated to be creative and to apply their detailed knowledge to the redesign of business processes.

Team Composition

Once organization-wide commitment has been secured from all departments involved in the reengineering effort and at different levels, the critical step of selecting a BPR team must be taken. This team will form the nucleus of the BPR effort, make key decisions and recommendations, and help communicate the details and benefits of the BPR program to the entire organization. The determinants of an effective BPR team may be summarized as follows:

- competency of the members of the team, their motivation,
- their credibility within the organization and their creativity,
- team empowerment, training of members in process mapping and brainstorming techniques,
- effective team leadership,
- proper organization of the team,
- complementary skills among team members, adequate size, interchangeable accountability, clarity of work approach, and
- specificity of goals.

The most effective BPR teams include active representatives from the following work groups: top management, business area responsible for the process being addressed, technology groups, finance, and members of all ultimate process users' groups. Team members who are selected from each work group within the organization will affect the outcome of the reengineered process according to their desired requirements. The BPR team should be mixed in depth and knowledge. For example, it may include members with the following characteristics:

- Members who do not know the process at all.
- Members who know the process inside-out.
- Customers, if possible.
- Members representing affected departments.
- One or two members of the best, brightest, passionate, and committed technology experts.
- Members from outside of the organization

Moreover, Covert (1997) recommends that in order to have an effective BPR team, it must be kept under ten players. If the organization fails to keep the team at a manageable size, the entire process will be much more difficult to execute efficiently and effectively. The efforts of the team must be focused on identifying breakthrough opportunities and designing new work steps or processes that will create quantum gains and competitive advantage.

Business Needs Analysis

Another important factor in the success of any BPR effort is performing a thorough business needs analysis. Too often, BPR teams jump directly into the technology without first assessing the current processes of the organization and determining what exactly needs reengineering. In this analysis phase, a series of sessions should be held with process owners and stakeholders, regarding the need and strategy for BPR. These sessions build a consensus as to the vision of the ideal business process. They help identify essential goals for BPR within each department and then collectively define objectives for how the project will affect each work group or department on individual basis and the business organization as a whole. The idea of these sessions is to conceptualize the ideal business process for the organization and build a business process model. Those items that seem unnecessary or unrealistic may be eliminated or modified later on in the diagnosing stage of the BPR project. It is important to acknowledge and evaluate all ideas in order to make all participants feel that they are a part of this important and crucial process. Results of these meetings will help formulate the basic plan for the project.

This plan includes the following:

- identifying specific problem areas,
- solidifying particular goals, and
- defining business objectives.

The business needs analysis contributes tremendously to the re-engineering effort by helping the BPR team to prioritize and determine where it should focus its improvements efforts.

The business needs analysis also helps in relating the BPR project goals back to key business objectives and the overall strategic direction for the organization. This linkage should show the thread from the top to the bottom of the organization, so each person can easily connect the overall business direction with the re-engineering effort. This alignment must be demonstrated from the perspective of financial performance, customer service, associate value, and the vision for the organization. Developing a business vision and process objectives relies, on the one hand, on a clear understanding of organizational strengths, weaknesses, and market structure, and on the other, on awareness and knowledge about innovative activities undertaken by competitors and other organizations.

BPR projects that are not in alignment with the organization's strategic direction can be counterproductive. There is always a possibility that an organization may make significant investments in an area that is not a core competency for the company and later outsource this capability. Such reengineering initiatives are wasteful and steal resources from other strategic projects. Moreover, without strategic alignment, the organization's key stakeholders and sponsors may find themselves unable to provide the level of support the organization needs in terms of resources, especially if there are other more critical projects to the future of the business, and are more aligned with the strategic direction.

Adequate IT Infrastructure

Researchers consider adequate IT infrastructure reassessment and composition as a vital factor in successful BPR implementation. Hammer (1990) prescribes the use of IT to challenge the assumptions inherent in the work process that have existed since long before the advent of modern computer and communications technology. Factors related to IT infrastructure have been increasingly considered by many researchers and practitioners as a vital component of successful BPR efforts.

- Effective alignment of IT infrastructure and BPR strategy,
- building an effective IT infrastructure,
- adequate IT infrastructure investment decision,
- adequate measurement of IT infrastructure effectiveness,
- proper information systems (IS) integration,
- effective reengineering of legacy IS,
- increasing IT function competency, and
- effective use of software tools are the most important factors that contribute to the success of BPR projects.

These are vital factors that contribute to building an effective IT infrastructure for business processes. BPR must be accompanied by strategic planning which addresses leveraging IT as a competitive tool. An IT infrastructure is made up of physical assets, intellectual assets, shared services, and their linkages. The way in which the IT infrastructure components are composed and their linkages determines the extent to which information resources can be delivered. An effective IT infrastructure composition process follows a top-down approach, beginning with business strategy and IS strategy and passing through designs of data, systems, and computer architecture.

Linkages between the IT infrastructure components, as well as descriptions of their contexts of interaction, are important for ensuring integrity and consistency among the IT infrastructure components. Furthermore, IT standards have a major role in reconciling various infrastructure components to provide shared IT services that are of a certain degree of effectiveness to support business process applications, as well as to guide the process of acquiring, managing, and utilizing IT assets. The IT infrastructure shared services and the human IT infrastructure components, in terms of their responsibilities and their needed expertise, are both vital to the process of the IT infrastructure composition. IT strategic alignment is approached through the process of integration between business and IT strategies, as well as between IT and organizational infrastructures.

Most analysts view BPR and IT as irrevocably linked. Walmart, for example, would not have been able to reengineer the processes used to procure and distribute mass-market retail goods without IT. Ford was able to decrease its headcount in the procurement department by 75 percent by using IT in conjunction with BPR, in another well-known example. The IT infrastructure and BPR are interdependent in the sense that deciding the information requirements for the new business processes determines the IT infrastructure constituents, and a recognition of IT capabilities provides alternatives for BPR. Building a responsive IT infrastructure is highly dependent on an appropriate determination of business process information needs. This, in turn, is determined by the types of activities embedded in a business process, and their sequencing and reliance on other organizational processes.

Effective Change Management

Al-Mashari and Zairi (2000) suggest that BPR involves changes in people behavior and culture, processes, and technology. As a result, there are many factors that prevent the effective implementation of BPR and hence restrict innovation and continuous improvement. Change management, which involves all human and social related changes and cultural adjustment techniques needed by management to facilitate the insertion of newly designed processes and structures into working practice and to deal effectively with resistance, is considered by many researchers to be a crucial component of any BPR effort. One of the most overlooked obstacles to successful BPR project implementation is resistance from those whom implementers believe will benefit the most. Most projects underestimate the cultural effect of major process and structural change and as a result, do not achieve the full potential of their change effort. Many people fail to understand that change is not an event, but rather a management technique.

Change management is the discipline of managing change as a process, with due consideration that employees are people, not programmable machines. Change is implicitly driven by motivation which is fueled by the recognition of the need for change. An important step towards any successful reengineering effort is to convey an understanding of the necessity for change. It is a well-known fact that organizations do not change unless people change; the better change is managed, the less painful the transition is.

Organizational culture is a determining factor in successful BPR implementation. Organizational culture influences the organization's ability to adapt to change. Culture in an organization is a self-reinforcing set of beliefs, attitudes, and behavior. Culture is one of the most resistant elements of organizational behavior and is extremely difficult to change. BPR must consider current culture in order to change these beliefs, attitudes, and behaviors effectively. Messages conveyed from management in an organization continually enforce current culture. Change is implicitly driven by motivation which is fueled by the recognition of the need for change.

The first step towards any successful transformation effort is to convey an understanding of the necessity for change. Management rewards system, stories of company origin and early successes of founders, physical symbols, and company icons constantly enforce the message of the current culture. Implementing BPR successfully is dependent on how thoroughly management conveys the new cultural messages to the organization. These messages provide people in the organization with a guideline to predict the outcome of acceptable behavior patterns. People should be the focus for any successful business change.

BPR is not a recipe for successful business transformation if it focuses on only computer technology and process redesign. In fact, many BPR projects have failed because they did not recognize the importance of the human element in implementing BPR. Understanding the people in organizations, the current company culture, motivation, leadership, and past performance is essential to recognize, understand, and integrate into the vision and implementation of BPR. If the human element is given equal or greater emphasis in BPR, the odds of successful business transformation increase substantially.

Ongoing Continuous Improvement

Many organizational change theorists hold a common view of organizations adjusting gradually and incrementally and responding locally to individual crises as they arise Common elements are:

- BPR is a successive and ongoing process and should be regarded as an improvement strategy that enables an organization to make the move from traditional functional orientation to one that aligns with strategic business processes.
- Continuous improvement is defined as the propensity of the organization to pursue incremental and innovative improvements in its processes, products, and services. The incremental change is governed by the knowledge gained from each previous change cycle.
- It is essential that the automation infrastructure of the BPR activity provides for performance measurements in order to support continuous improvements. It will need to efficiently capture appropriate data and allow access to appropriate individuals.
- To ensure that the process generates the desired benefits, it must be tested before it is deployed to the end users. If it does not perform satisfactorily, more time should be taken to modify the process until it does.

- A fundamental concept for quality practitioners is the use of feedback loops at every step of the process and an environment that encourages constant evaluation of results and individual efforts to improve.
- At the end user's level, there must be a proactive feedback mechanism that provides for and facilitates resolutions of problems and issues. This will also contribute to a continuous risk assessment and evaluation which are needed throughout the implementation process to deal with any risks at their initial state and to ensure the success of the reengineering efforts.
- Anticipating and planning for risk handling is important for dealing effectively with any risk when it first occurs and as early as possible in the BPR process. It is interesting that many of the successful applications of reengineering described by its proponents are in organizations practicing continuous improvement programs.
- Hammer and Champy (1993) use the IBM Credit Corporation as well as Ford and Kodak, as examples of companies that carried out BPR successfully due to the fact that they had long-running continuous improvement programs.

In conclusion, successful BPR can potentially create substantial improvements in the way organizations do business and can actually produce fundamental improvements for business operations. However, in order to achieve that, there are some key success factors that must be taken into consideration when performing BPR.

BPR success factors are a collection of lessons learned from reengineering projects and from these lessons common themes have emerged. In addition, the ultimate success of BPR depends on the people who do it and on how well they can be committed and motivated to be creative and to apply their detailed knowledge to the reengineering initiative. Organizations planning to undertake BPR must take into consideration the success factors of BPR in order to ensure that their reengineering related change efforts are comprehensive, well-implemented, and have minimum chance of failure. This has been very beneficial in all terms

Critique

Many companies used reengineering as a pretext to downsizing, though this was not the intent of reengineering's proponents; consequently, reengineering earned a reputation for being synonymous with downsizing and layoffs.

In many circumstances, reengineering has not always lived up to its expectations. Some prominent reasons include:

- Reengineering assumes that the factor that limits an organization's performance is the ineffectiveness of its processes (which may or may not be true) and offers no means of validating that assumption.

- Reengineering assumes the need to start the process of performance improvement with a "clean slate," i.e. totally disregard the *status quo*.
- According to Eliyahu M. Goldratt (and his Theory of Constraints) reengineering does not provide an effective way to focus improvement efforts on the organization's constraint.

Others have claimed that reengineering was a recycled buzzword for commonly-held ideas. Abrahamson (1996) argued that fashionable management terms tend to follow a lifecycle, which for Reengineering peaked between 1993 and 1996 (Ponzi and Koenig 2002). They argue that Reengineering was in fact nothing new (as e.g. when Henry Ford implemented the assembly line in 1908, he was in fact reengineering, radically changing the way of thinking in an organization).

The most frequent critique against BPR concerns the strict focus on efficiency and technology and the disregard of people in the organization that is subjected to a reengineering initiative. Very often, the label BPR was used for major workforce reductions. Thomas Davenport, an early BPR proponent, stated that:

"When I wrote about "business process redesign" in 1990, I explicitly said that using it for cost reduction alone was not a sensible goal. And consultants Michael Hammer and James Champy, the two names most closely associated with reengineering, have insisted all along that layoffs shouldn't be the point. But the fact is, once out of the bottle, the reengineering genie quickly turned ugly."

Hammer similarly admitted that:

"I wasn't smart enough about that. I was reflecting my engineering background and was insufficient appreciative of the human dimension. I've learned that's critical."

Outsourcing

In business, outsourcing involves the contracting out of a business process (e.g. payroll processing, claims processing) and operational, and/or non-core functions (e.g. manufacturing, facility management, call center support) to another party. The concept "outsourcing" came from the American Glossary 'outside resourcing' and it dates back to at least 1981. Outsourcing sometimes involves transferring employees and assets from one firm to another, but not always. Outsourcing is also the practice of handing over control of public services to for-profit corporations.

Outsourcing includes both foreign and domestic contracting, and sometimes includes offshoring (relocating a business function to a distant country) or nearshoring (transferring a business process to a nearby country). Many people confuse outsourcing and offshoring – however they are different. A company can outsource (work with a service provider) and not offshore to a distant country. For example, in 2003 Procter and Gamble outsourced their facilities management support, but it did not involve offshoring. Financial savings from lower international labor rates can provide a major motivation for outsourcing or offshoring. There can be tremendous savings from lower international labor rates when offshoring.

The opposite of outsourcing, insourcing, entails bringing processes handled by third-party firms in-house, and is sometimes accomplished via vertical integration. However, a business can provide a contract service to another organization without necessarily insourcing that business process.

Overview

Two organizations may enter into a contractual agreement involving an exchange of services, expertise, and payments. Outsourcing is said to help firms to perform well in their core competencies, fuel innovation, and mitigate a shortage of skill or expertise in the areas where they want to outsource.

In the early 21st century, businesses increasingly outsourced to suppliers outside their own country, sometimes referred to as offshoring or offshore outsourcing. Several related terms have emerged to refer to various aspects of the complex relationship between economic organizations or networks, such as nearshoring, crowdsourcing, multisourcing, strategic alliances/strategic partnerships, strategic outsourcing., and Vested outsourcing.

Outsourcing can offer greater budget flexibility and control. Outsourcing allows organizations to pay for the services and business functions they need, when they need them. It also reduces the need to hire and train specialized staff, brings in fresh engineering expertise, and can reduce capital and operating expenses and risk.

"Do what you do best and outsource the rest" has become an internationally recognized business tagline first "coined and developed" in the 1990s by the "legendary management consultant" Peter Drucker. The slogan was primarily used to advocate outsourcing as a viable business strategy. It has been said that Mr. Drucker began explaining the concept of "Outsourcing" as early as 1989 in his Wall Street Journal (WSJ) article entitled "Sell the Mailroom."

From Drucker's perspective, a company should only seek to subcontract in those areas in which it demonstrated no special ability. The business strategy outlined by his slogan recommended that companies should take advantage of a specialist provider's knowledge and economies of scale to improve performance and achieve the service needed.

In 2009 by way of recognition, Peter Drucker posthumously received a significant honor, when he was inducted into the Outsourcing Hall of Fame for his outstanding work in the field.

Reasons for Outsourcing

Companies primarily outsource to reduce certain costs – such as peripheral or "non-core" business expenses, high taxes, high energy costs, excessive government regulation/mandates, production and/or labor costs. The incentive to outsource may be greater for U.S. companies due to unusually high corporate taxes and mandated benefits, like social security, Medicare, and safety protection (OSHA regulations). At the same time, it appears U.S. companies do not outsource to reduce executive or managerial costs. For instance, executive pay in the United States in 2007 was more than 400 times more than average workers—a gap 20 times bigger than it was in 1965.

Digital Outsourcing

Many think of outsourcing as it relates to manufacturing (e.g. the "made in China" label on the product you buy). However, outsourcing of white collar work has grown rapidly since the early 21st century. The digital workforce of countries like India and China are only paid a fraction of what would be minimum wage in the US. On average, software engineers are getting paid between 250,000 and 1,500,000 rupees ($4,000 to $23,000) in India as opposed to the $40,000-$100,000 in countries like US and Canada. Outsourcing has also expanded to include many different countries; Costa Rica has become a big source for outsourcing work as it offers the advantage of a highly educated labor force, a large bilingual population, stable democratic government, shares similar time zones with the United States, and it takes only a few hours to travel between Costa Rica and the US. Companies such as Intel, Procter & Gamble, HP, Gensler, Amazon and Bank of America have big operations in Costa Rica. In the recent years there has been an exponential growth in white collar work with service providers emerging in a wide range of activities, from banking and legal services to companies like Resourcesus US, a pioneer in outsourcing services for architecture firms in United States. Unlike outsourced manufacturing, outsourced white collar work, offers workers the flexibility to choose their working hours, and which companies to work for. With many individuals telecommuting from home, the companies that require this type of work do not need to allocate additional funds for setting up of office space, management salary, and employee benefits as these individuals are contracted workers.

Implications

Management Processes

Globalization and complex supply chains, along with greater physical distance between higher management and the production-floor employees often requires a change in management methodologies, as inspection and feedback may not be as direct and frequent as in internal processes. This often requires the assimilation of new communication methods such as voice over IP, instant messaging, and Issue tracking systems, new time management methods such as time tracking software, and new cost- and schedule-assessment tools such as cost estimation software.

Communications and Customer Service

In the area of call center outsourcing, organizations that are not experienced in working with outsourced call centers may suffer from lower end-user-experience as a result of outsourcing. This is exacerbated when outsourcing is combined with offshoring in regions where the first language and culture are different.

For example, foreign call center agents may speak with different linguistic features such as accents, word use and phraseology, which may impede comprehension. The visual cues that are missing in a telephone call may lead to misunderstandings and difficulties.

Governance

In 1979, Nobel laureate Oliver Williamson wrote that the governance structure is the "framework within which the integrity of a transaction is decided." Adding further that "because contracts are varied and complex, governance structures vary with the nature of the transaction." University of Tennessee researchers have been studying complex outsourcing relationships since 2003. Emerging thinking regarding strategic outsourcing is focusing on creating a contract structure in which the parties have a vested interest in managing what are often highly complex business arrangements in a more collaborative, aligned, flexible, and credible way.

Security

Before outsourcing, an organization is responsible for the actions of their entire staff, sometimes a substantial liability. When these same people are transferred to an outsource service provider, they may not even change desks. But their legal status changes. They are no longer directly employed by (and responsible to) the organization. This creates legal, security and compliance issues that are often addressed through the contract between the client and the suppliers. This is one of the most complex areas of outsourcing and sometimes involves a specialist third-party adviser.

Fraud is a specific security issue as well as criminal activity, whether it is by employees or the supplier staff. However, it can be disputed that fraud is more likely when outsourcing is involved. For example, in April 2005, a high-profile case involved the theft of $350,000 from four Citibank customers when call-center workers acquired the passwords to customer accounts and transferred the money to their own accounts opened under fictitious names. Citibank did not find out about the problem until the American customers noticed discrepancies with their accounts and notified the bank.

Insourcing

Outsourcing has gone through many iterations and reinventions. Some outsourcing contracts have been partially or fully reversed, citing an inability to execute strategy, lost transparency & control, onerous contractual models, a lack of competition, recurring costs, hidden costs, and so on. Companies shifting to insourcing often cite the desire to increase control, compliance and to gain competitive differentiation through vertical integration or the development of shared services, commonly called a center of excellence.

Further, the label *outsourcing* has been found to be used for too many different kinds of exchanges often in confusing ways. For example, global software development, which often involves people working in different countries, cannot simply be called outsourcing. The outsourcing-based market model fails to explain why these development projects are jointly developed, and not simply bought and sold in the marketplace. Recently, a study has identified an additional system of governance, termed algocracy, which appears to govern global software projects alongside bureaucratic and market-based mechanisms. The study distinguishes code-based governance system from bureaucracy and the market, and underscores the prominent features of each organizational form in terms of its ruling mechanism: bureaucracy (legal-rational), the market (price), and algocracy (programming or algorithm). So, global software development projects, though not insourced, are not outsourced either. They are in-between, in a process that is sometimes termed "remote in-sourcing." Projects are developed together where a common software platform allows different teams around the world to work on the same project together.

Standpoint of Labor

From the standpoint of labor, outsourcing may represent a new threat, contributing to worker insecurity, and is reflective of the general process of globalization and economic polarization.

On June 26, 2009, Jeff Immelt, the CEO of General Electric, called for the United States to increase its manufacturing base employment to 20% of the workforce, commenting that the U.S. has outsourced too much and can no longer rely on consumer spending to drive demand.

Standpoint of Government

Western governments may attempt to compensate workers affected by outsourcing through various forms of legislation. In Europe, the Acquired Rights Directive attempts to address the issue. The Directive is implemented differently in different nations. In the United States, the Trade Adjustment Assistance Act is meant to provide compensation for workers directly affected by international trade agreements. Whether or not these policies provide the security and fair compensation they promise is debatable.

Policy-making Strategy

A main feature of outsourcing influencing policy-making is the unpredictability it generates regarding the future of any particular sector or skill-group. The uncertainty of future conditions influences governance approaches to different aspects of long-term policies.

Competitiveness strategy

Economic growth requires change, therefore a governance disposed to helping social and economic structures adapt to the changing environment will facilitate growth and a stable transition to new economic structures., until the economic structures become detrimental to the social, political and cultural structures. In less economically developed countries, policies which embrace the global phenomenon of outsourcing are a logical response to the ongoing movement towards "open markets" and "trade liberalization." Outsourcing, when interpreted as a trade phenomenon, complements trade liberalization strategies not only by promoting technological spillovers and capital inflows but also by offsetting the increasing levels of unemployment which result from opening up domestic markets. As prices adjust to those in the global market they no longer reflect domestic productivity, driving lower-productivity firms in the previously protected sectors out of business. Economic theorists argue that the resulting unemployment is only temporary as workers readjust and are eventually incorporated into the country's most productive sectors, namely those which enjoy a competitive edge over other players in the international market. Nonetheless, rapid liberalization of markets in developing nations has not maximized the productivity potential of the region. In the Global South, where technological development is drastically lower than in the North, the redeployment of human and capital resources into new export markets has not come at the cost of necessarily low-productivity sectors but rather underdeveloped ones. In other words, many of the previously protected sectors were not competitive yet on a global scale, not because they naturally lacked the comparative advantage, but because industry efficiency had not yet been reached. In such cases where liberalization stunts the growth of potential industries, unemployment is a reflection of many underemployed resources. Outsourcing fills in the gap of receding protected national industries, improving employment and living standards. Among other economic externalities, outsourcing promotes capital inflows and infrastructure. In Mexico, wage convergence was faster in cities where outsourcing first took hold through maquiladoras, along the US-Mexican border. Studies suggest that for every 10% increase in US wages, northern cities in Mexico which are most influenced by outsourcing would experience wage rises of 2.5%, about 0.69% higher than in inner cities. Corruption and reduced tax revenues after signing the NAFTA Treaty have limited the economic resources available to the Mexican government, thus explaining the difference in investment policies between Mexico and China. Conversely, one of the successes of Asian countries in the twentieth century has been their promotion of higher rates of saving and investment. Studies suggest that the increase in capital input fueled the 'Asian miracle' rather than improvements in productivity and industrial efficiency. Though the previous conclusion suggests production conditions in the region remained static, the situation in East Asia experienced rapid transformations. Not only were national educational rates raised drastically, but there was also an increase in patenting and research and development expenditures. Rising levels of education, urbanization and even of patenting illustrate the active role of the government in advancing education as well as encouraging research and development.

Education strategy

Jobs become outsourced not based on the skill-level group it represents, but rather based on a variety of other factors including transportation cost of ideas, wage and labour productivity edge. Because of the overall uncertainty regarding the future dynamics of outsourcing it is not possible to predict the nature of labour demand in different regions. To better prepare the domestic workforce to future industry demands, therefore, national education programs ought to focus on flexibility and diversity of skills rather than on any specific task-oriented skills. Emphasis should go on preparing students both to succeed in non-habitual tasks and to adapt to changes in labour demands in the market. A specific goal that ought to be adopted is teaching students how to learn rather than teach them particular skills. This strategy would help students adapt to changing skill requirements in the future thus reducing friction from structural unemployment.

Welfare state strategy

The uncertainty regarding the domestic productivity edge renders caution a key element of governance to ensure a sustainable regional developmenloyed re-enter the work forcc and smoothly transition into high-demand labour opportunities – potentially through re-training programs –, the government should also address the socio-economic struggle and other welfare concerns of displaced employees. Negative welfare effects of outsourcing have gathered substantial public attention. The possibility of outsourcing has internationalized labour markets which used to be local, opening up jobs which were traditionally non-traded to international competition. The resulting combination of lower wages and unemployment for certain jobs has driven the perceived 'losers' to engage in heated political debate. Labour unions in the European Union have succeeded in pushing through protectionist policies in favour or lower-skilled groups throughout the 1970s and 1980s, including the Common Agricultural Policy on farming. Interest groups opposing outsourcing have been more active to voice their disapproval because the negative outcomes of the phenomenon are more concentrated on specific groups of people, namely those losing jobs to external competition, whereas the benefits from it become dissipated among the population at large. Overall lower prices and greater quality and variety of goods in domestic markets are some of the benefits of exploiting a country's comparative advantage through outsourcing. Unlike the alleged 'losers' from outsourcing, those affected positively by it lack the motivation to organize to voice their support. There has been a wave of protectionism concerned with deep changes in the social structure allegedly imposed on the global system through globalization and outsourcing. The activists see a readjustment of class systems and highlight an increased fracture in societies between the 'haves' and the 'have-nots' as different groups adjust to increasingly or decreasingly advantageous positions in the system of outsourcing. Opponents of outsourcing have also denounced it as a threat to local cultural integrity. The argument on cultural disintegration points to the standardization of practices and norms as multinational corporations become involved with industries in regions culturally different from those in the country of origin. The alleged diffusion of culture has raised concern over the endurance of cultural norms and values, sociopolitical institutions and frameworks, or even cultural preferences and traditions in a context of increasing foreign presence. Increased uncertainty regarding future socioeconomic security ought to be met with policies promoting equality and a fair redistribution of economic gains for a government to maintain its voters' favour. Because of overall unpredictability, governments will likely need to reassure civilians that the burden of employment jobs resulting from outsourcing will be shared among taxpayers. The fluctuations in employment levels are determined by the types of jobs which can be profitably outsourced or offshored. Domestic jobs become offshored or outsourced when lower productivity in other regions is compensated by lower wages, making outsourcing profitable even despite the added costs of transportation. The overall cost-effectiveness of the spatial unbundling of the industrial process thus depends on the cost of transporting specific services or ideas given the available technology. Because of this reason technological advancements such as the telecommunications revolution, air shipping or the Internet have deeply accelerated outsourcing and may continue to boost this process. The future results of technological ingenuity and innovation are unknown, as are its potential impacts employment levels on any given task or job across regions. In the Global South, policies attracting multinational corporations can help increase employment levels and promote growth. Governments which pursue such strategies facilitate welfare protection given the context of increased unemployment in industries which cannot compete with the international market due to trade liberalization policies.

Industrial policy

Outsourcing results from an internationalization of labor markets as more tasks become tradable. According to leading economist Greg Mankiw, the labour market functions under the same forces as the market of goods, with the underlying implication that the greater the number of tasks available to being moved, the better for efficiency under the gains from trade. With technological progress, more tasks can be offshored at different stages of the overall corporate process.

Environmental policy

There are widespread claims that outsourcing has pushed environmental standards down in developing regions as countries compete to attract foreign projects and investment. Similar to lower wages, lower health and environmental regulations contribute to giving a country a comparative advantage over another due to lower production costs. The controversy this raises, however, is that unlike wages, lower health or environmental standards does benefit the new employees joining the workforce. Import competition has caused a de facto 'race-to-the-bottom' where countries lower environmental regulations to secure a competitive edge for their industries relative to other countries. As Mexico competes with China over Canadian and American markets, its national Commission for Environmental Cooperation has not been active in enacting or enforcing regulations to prevent environmental damage from increasingly industrialized Export Processing Zones. Similarly, since the signing of NAFTA heavy industries have increasingly moved to the US which has a comparative advantage due to its abundant presence capital and well-developed technology. A further example of environmental de-regulation with the objective of protecting trade incentives have been the numerous exemptions to carbon taxes in European countries during the 1990s. The evidence suggests that even if outsourcing has promoted lower environmental protection, there are no intrinsic geographic implications that the Global South has been more negatively affected than the North. o Although outsourcing can influence environmental de-regulatory trends, the added cost of preventing pollution does not majorly determine trade flows or industrialization.

Globalization and Socio-economic Implications

Global Inequality and Development

Industrialization

Outsourcing has contributed to further levelling of global inequalities as it has led to general trends of industrialization in the Global South and deindustrialization in the Global North.

Even though outsourcing has promoted a movement of industrial sites from the Global North to Global South regions, it has not been the only reason for the concurrent deindustrialization and industrialization of the North and South respectively. Deindustrialization in more economically and technologically developed regions has also been affected by increased industrial productivity.

The rise in industrial efficiency which characterized development in developed countries has occurred as a result of labor-saving technological improvements. Although these improvements do not directly reduce employment levels but rather increase output per unit of work, they can indirectly diminish the amount of labor required for fixed levels of output. Likewise, a trending shift in demand towards non-tradable services such as those in the health-care or government sectors has further accelerated deindustrialization in the Global North. Since these tasks cannot be outsourced, the demand for them needs to be met domestically abiding by the local market price. Consequently, a shift in the labor force towards fulfilling these profitable services has mostly taken place at the expense of industry since the agricultural sector in the early industrializing Global North had already been maximizing its labour capacity.

Despite the variety of domestic and international factors affecting deindustrialization in the Global North, those concerning the external influence of the global market have been the most influential ones since 1994.

The recent industrialization process outsourcing has encouraged in the Global South has taken place at a much faster pace than it did during its beginnings in the North, given that the well-developed technology was already developed, and merely spread to further regions.

Growth and income

The almost simultaneous industrialization of the developing Global South and deindustrialization of the more developed Global North resulted in an international convergence of incomes.

Outsourcing has been characterized by rapid rates of growth and industrialization in the Global South. Conversely, the Global North has experienced a moderate slowdown in growth. Patterns of global industrialization and deindustrialization are explained by a combination of models in economic geography and economic growth. Models in economic geography illustrate that decreasing communication costs reduce the spatial clustering of industrial development. The lower cost to the spread of ideas improved coordination and communication within corporations across greater geographic distances. This process originated roughly after the international chaos of the World Wars, as a consequence of advancements in information technology during the 1960s. Further adaptations to technological progress, particularly the spread of the internet and liberalization of the telecommunications industry, permitted an acceleration of the movement of ideas and consequently of outsourcing since about the 1980s.

Urbanization

The pace of urbanization in the Global North decreased considerably relative to its high levels following the Industrial Revolution. Rates of urban growth have been higher in the Global South.

Trade

Outsourcing emerged with a new wave of globalization marked by high volumes of trade and capital flows. The increased movement of capital and goods contrasted starkly with the barriers and protectionism prevalent throughout the World Wars and the Great Depression during the Interwar Years.

Migration

The level of migration has remained relatively low, particularly compared to the mass migratory trends which characterized the Industrial Revolution roughly between 1850 and 1914., probably because labor markets are not free now. Countries now have discrimination labor laws, only allow people with citizenship cards live and work free in their territories, even getting a citizenship card is difficult for some one not born in their territory. Free labor markets, discrimination based with a person skills would help reduce outsourcing problems, letting people freely follow their jobs in other countries.

Domestic Inequality

Outsourcing in the Global North led to a rising divergence in domestic employment levels in different tasks within a given industry, making tackling unemployment more difficult for governments as job losses ceased to be primarily sectoral.

By Country

United States

"Outsourcing" has become a continuing political issue in the United States, having been conflated with offshoring, during the 2004 U.S. presidential election. The political debate centered on outsourcing's consequences for the domestic U.S. workforce. Democratic U.S. presidential candidate John Kerry criticized U.S. firms that outsource jobs abroad or that incorporate overseas in tax havens to avoid paying their "fair share" of U.S. taxes during his 2004 campaign, calling such firms "Benedict Arnold corporations".

Criticism of outsourcing, from the perspective of U.S. citizens, generally revolves around the costs associated with transferring control of the labor process to an external entity in another country. A Zogby International poll conducted in August 2004 found that 71% of American voters believed that "outsourcing jobs overseas" hurt the economy while another 62% believed that the U.S. government should impose some legislative action against companies that transfer domestic jobs overseas, possibly in the form of increased taxes on companies that outsource. One prediction (from 2010) claims that, by 2014, more than 1.3 million positions will disappear because of "the accelerated movement of work to India ..." and some other countries willing to accept outsourced jobs. President Obama promoted an act titled 'Bring Jobs Home Act' that would help reshore jobs by giving incentives such as a tax cut or a 20 percent tax credit for moving operations back to the USA. The same bill was reintroduced in the 113th United States Congress as the Bring Jobs Home Act (S. 2569; 113th Congress).

Union busting is one possible cause of outsourcing. As unions are disadvantaged by union busting legislation, workers lose bargaining power and it becomes easier for corporations to fire them and ship their job overseas.

Another rationale is the high corporate income tax rate in the U.S. relative to other OECD nations, and the practice of taxing revenues earned outside of U.S. jurisdiction, a very uncommon practice. However, outsourcing is not solely a U.S. phenomenon as corporations in various nations with low tax rates outsource as well, which means that high taxation can only partially, if at all, explain US outsourcing. For example, the amount of corporate outsourcing in 1950 would be considerably lower than today, yet the tax rate was actually higher in 1950.

It is also suggested that lowering the corporate income tax and ending the double-taxation of foreign-derived revenue (taxed once in the nation where the revenue was raised, and once from the U.S.) will alleviate corporate outsourcing and make the U.S. more attractive to foreign companies. However, while the US has a high official tax rate, the actual taxes paid by US corporations may be considerably lower due to the use of tax loopholes, tax havens, and "gaming the system". Rather than avoiding taxes, outsourcing may be mostly driven by the desire to lower labor costs. Sarbanes-Oxley has also been cited as a factor for corporate flight from U.S. jurisdiction.

European Union

Where outsourcing involves the transfer of an undertaking, it is subject to Council Directive 77/187 of 14 February 1977, on the approximation of the laws of the Member States relating to the safeguarding of employees' rights in the event of transfers of undertakings, businesses or parts of businesses (as amended by Directive 98/50/EC of 29 June 1998; consolidated in Directive 2001/23 of 12 March 2001). Under that directive, rights acquired by employees with the former employer are to be safeguarded when they, together with the undertaking in which they are employed, are transferred to another employer, i.e., the contractor. An example of a case involving such contracting-out was the decision of the European Court of Justice in Christel Schmidt v. Spar- und Leihkasse der früheren Ämter Bordesholm, Kiel und Cronshagen, Case C-392/92 [1994]. Although subsequent decisions have disputed whether a particular contracting-out exercise constituted a transfer of an undertaking (for example, Ayse Süzen v. Zehnacker Gebäudereinigung GmbH Krankenhausservice, Case C-13/95 [1997]), in principle, employees of an enterprise outsourcing part of its activities in which they are employed may benefit from the protection offered by the directive.

Seeking to implement the cost-cutting solutions, many Western European firms have been transferring tech projects eastward. For example, Deutsche Bank has some of its software developed in Ukraine, Siemens possess R&D center in Romania. Europe Outsourcing has produced outstanding results and henceforward they are increasing them in numbers.

Despite unfavorable economic conditions from 2007 to 2009, the outsourcing services market continued to flourish in Central and Eastern European. In 2008 when the inflow of investment in Western Europe was down by 48%, it fell by only 9% in Central and Eastern Europe. In Poland alone, during 2009, the year following the global economic downturn, approximately 10,000 jobs were created in business process outsourcing (BPO).

Co-sourcing

Co-sourcing is a business practice where a service is performed by staff from inside an organization and also by an external service provider. It can be a service performed in concert with a client's existing internal audit department. The scope of work may focus on one or more aspects of the internal audit function. Co-sourcing can serve to minimize sourcing risks, increase transparency, clarity and lend toward better control over the processes outsourced.

Examples of co-sourcing services are supplementing the in-house internal audit staff with specialized skills such as information risk management or integrity services, providing routine assistance to in-house auditing for operations and control evaluations in peak period activity and conducting special projects such as fraud investigation or plant investment appraisals. Another example of co-sourcing is outsourcing part of software development or software maintenance activities to an external organization, while keeping part of the development in-house. Other internal business activities such as HR and administrative tasks can also be co-sourced by employee leasing companies.

Identity Management Co-sourcing

It is an approach to enterprise identity management in which the identity service interacts directly or through some technical footprint with an organization's Information Technology (IT) identity backend infrastructure (directories, databases, and other identity repositories). The organization and the external service provider typically have a shared responsibility for building, hosting and operating the identity service. The balance of this responsibility can vary depending on the service levels required, and span from an all on-premises deployment, where the identity service is built, hosted and operated within the organization's IT infrastructure and managed on-premises by the external service provider. This contrasts with an "all in-the-cloud" service scenario, where the identity service is built, hosted and operated by the service provider in an externally hosted, cloud computing infrastructure.

Counterwave Outsourcing

Some firms have reversed the decision to outsource because the results were not entirely as expected. The circumstances which allow firms to unbundle the different tasks or stages of its manufacturing process into different locations have not been fully determined. Though the nature of the tasks plays a role determining their interconnectedness, other factors such as innovation in the manufacturing process or advances in transport and communication technology also affect the need for direct contact among employees. As the process which ties tasks together within firms remains unclear, there is a degree of uncertainty about which tasks need to remain geographically clustered together. In many cases firms took risks experimenting with outsourcing while lacking a firm understanding of the relationship among internal tasks and its spatial implications.

Despite saving money, companies have often faced unexpected drawbacks from outsourcing, such as miscommunication or lower quality of intermediate products, which end up delaying the overall production process. Another reason for a decrease in outsourcing is that many jobs that were subcontracted abroad have been replaced by technological advances.

According to a Deloitte Consulting survey carried out in 2005, a quarter of the companies which had outsourced tasks reversed their strategy. Many big companies like Lenovo considered turning around outsourcing strategies of outsourcing.

Public opinion in the US and other Western powers opposing outsourcing was particularly strengthened by the drastic increase in unemployment as a result of the 2007-2008 financial crisis. In the first decade from 2000 to 2010, the US experienced a net loss of 687,000 jobs due to outsourcing, primarily in the computers and electronics sector. Public disenchantment with outsourcing has not only stirred political responses, as seen in the 2012 US Presidential campaigns, but it has also made companies more reluctant to outsource or offshore jobs.

A more recent (2016) Deloitte survey suggested that companies are no longer reluctant to outsource. Deloitte's survey identified three trends:

- Companies are broadening their approach to outsourcing as they begin to view it as more than a simple cost-cutting play
- Organizations are "redefining the ways they enter into outsourcing relationships and manage the ensuing risks."
- Organizations are changing the way they are managing their relationships with outsourcing providers to "maximize the value of those relationships."

Business Activity Monitoring

Business activity monitoring (BAM) is software that aids in monitoring of business activities, as those activities are implemented in computer systems.

The term was originally coined by analysts at Gartner, Inc. and refers to the aggregation, analysis, and presentation of real-time information about activities inside organizations and involving customers and partners. A business activity can either be a business process that is orchestrated by business process management (BPM) software, or a business process that is a series of activities spanning multiple systems and applications. BAM is an enterprise solution primarily intended to provide a real-time summary of business activities to operations managers and upper management.

Goals and Benefits

The goals of business activity monitoring is to provide real time information about the status and results of various operations, processes, and transactions. The main benefits of BAM are to enable an enterprise to make better informed business decisions, quickly address problem areas, and re-position organizations to take full advantage of emerging opportunities.

Key Features

One of the most visible features of BAM solutions is the presentation of information on dashboards containing the key performance indicators (KPIs) used to provide assurance and visibility of activity and performance. This information is used by technical and business operations to provide visibility, measurement, and assurance of key business activities. It is also exploited by event correlation to detect and warn of impending problems.

Although BAM systems usually use a computer dashboard display to present data, BAM is distinct from the dashboards used by business intelligence (BI) insofar as events are processed in real-time or near real-time and pushed to the dashboard in BAM systems, whereas BI dashboards refresh at predetermined intervals by polling or querying databases. Depending on the refresh interval selected, BAM and BI dashboards can be similar or vary considerably.

Some BAM solutions additionally provide trouble notification functions, which allows them to interact automatically with the issue tracking system. For example, whole groups of people can be sent e-mails, voice or text messages, according to the nature of the problem. Automated problem solving, where feasible, can correct and restart failed processes.

Deployment Effort

In nearly all BAM deployments extensive tailoring to specific enterprises is required. Many BAM solutions seek to reduce extensive customization and may offer templates that are written to solve common problems in specific sectors, for example banking, manufacturing, and stockbroking. Due to the high degree of system integration required for initial deployment, many enterprises use experts that specialize in BAM to implement solutions.

BAM is now considered a critical component of Operational Intelligence (OI) solutions to deliver visibility into business operations. Multiple sources of data can be combined from different organizational silos to provide a common operating picture that uses current information. Wherever real-time insight has the greatest value, OI solutions can be applied to deliver the needed information.

Processing Events

All BAM solutions process events. While most of the first BAM solutions were closely linked to BPM solutions and therefore processed events emitted as the process was being orchestrated, this had the disadvantage of requiring enterprises to invest in BPM before being able to acquire and use BAM. The newer generation of BAM solutions are based on complex event processing (CEP) technology, and can process high volumes of underlying technical events to derive higher level business events, therefore reducing the dependency on BPM, and providing BAM to a wider audience of customers.

Examples

A bank might be interested in minimizing the amount of money it borrows overnight from a central bank. Interbank transfers must be communicated and arranged through automation by a set time each business day. The failure of any vital communication could cost the bank large sums in interest charged by the central bank. A BAM solution would be programmed to become aware of each message and await confirmation. Failure to receive confirmation within a reasonable amount of time would be grounds for the BAM solution to raise an alarm that would set in motion manual intervention to investigate the cause of the delay and to push the problem toward resolution before it becomes costly.

Another example involves a mobile telecommunications company interested in detecting a situation whereby new customers are not set up promptly and correctly on their network and within the various CRM and billing solutions. Low-level technical events such as messages passing from one application to another over a middleware system, or transactions detected within a database logfile, are processed by a CEP engine. All events relating to an individual customer are correlated in order to detect an anomalous situation whereby an individual customer has not been promptly or correctly provisioned, or set up. An alert can be generated to notify technical operations or to notify business operations, and the failed provisioning step may be restarted automatically.

Collective Business System

A collective business system or collective business model is a business organization or association typically composed of relatively large numbers of businesses, tradespersons or professionals in the same or related fields of endeavor, which pools resources, shares information or provides other benefits for their members. In the past, collective business systems such as the trade association, the cooperative and the franchise were created to allow groups of independently owned businesses with common interests to successfully compete in the marketplace.

Overview

Beginning in the middle of the twentieth century, the business world in the United States of America and many European countries has witnessed the consolidation of all types of businesses through mergers, rollups or acquisitions. Typically, by the end of the consolidation process, a particular industry or profession becomes dominated by three or four nationally based enterprises. Under these circumstances many smaller companies (often serving only local requirements and in private ownership) are often forced out of business or decide to sell to one of the dominant entities because it can no longer compete profitably with them. Typically, locally based businesses are unable to compete because they lack the capital, global marketing capabilities, purchasing power and expensive technology necessary to operate efficiently. This trend toward consolidation is expected to continue well into the twenty-first century and is sometimes shortened in developing countries, where the initial business in a market sector may be a single, large enterprise able to compete on the international market.

Presently, and for the foreseeable future, the inherent limitations of traditional collective business systems such as the trade organization, the cooperative and the franchise render them considerably less effective than they once were in advancing the business interests of their constituents. The following is a brief synopsis of those traditional collective business systems.

Trade Association

Trade associations are non-profit organizations in which the individual members are companies or individuals engaged in a common business pursuit. Competitors join together to create a platform format in which they deal with common problems of their industry. Any applicant meeting the standards of the association must be accepted as a member. Anti-trust law prohibits a member trade association from denying an otherwise qualified applicant's membership based upon a geographical proximity to an existing member. Trade associations commonly offer their members educational programs, the opportunity to come together at meetings to discuss common problems, and marketing materials designed to be imprinted by each member with its relevant information. Trade associations also offer elective group purchasing plans. The trade association bears no credit risk in these transactions but instead, provides chosen vendors with access to a large body of member customers. Because the trade association does not pledge its credit, the vendor must rely upon the credit worthiness of each purchaser.

To sustain its operations, a trade association generally receives an initiation fee and/

or a yearly membership fee (collectively "dues") from its members, and it may collect rebates or commissions from the purchasing plan suppliers.

The trade organization imposes relatively low membership dues on its members. However, because initiation fees and annual membership fees are nominal, the trade organization lacks the ability to engage in offering its members national marketing capability, access to expensive technologies and cost-effective purchasing programs for major purchases due to a lack of capital. Furthermore, being non-profit, trade associations do not have the management mentality necessary to sustain major projects such as national sales and marketing.

While offering potentially valuable services to businesses, few trade associations offer much direct help in the major business areas of purchasing, production or marketing.

Cooperative

A cooperative is a non-profit organization somewhat similar to a trade association. A significant difference between the cooperative and the trade association, however, is that with a trade association, the members have a non-equity position in the association, whereas in the typical cooperative the members will have an equity interest as all members of the cooperative own a portion of the cooperative.

The cooperative utilizes its volume leverage with suppliers in purchasing products and services for less than the individual member company could obtain outside of the cooperative. The cooperative marks up the purchased products or services in order to cover operating expenses. Any net income achieved by the cooperative is then returned to the cooperative members in the form of a redistribution of profits or dividends.

There are also cooperatives in which the sole function is for marketing and advertising in a given region. New car dealers and fast food franchisees typically form marketing and advertising cooperatives.

Franchise

The employees typically are required to wear uniforms and to dress as specified by the franchisor. Franchises can be offered by the franchisor on a territorial basis without violating antitrust laws. Ordinarily, the franchisee owns the non-real estate assets of a franchise. There is generally a substantial fee paid by the franchisee for the privilege of becoming a franchisee.

Internet of Things

The Internet of things (stylised Internet of Things or IoT) is the internetworking of physical devices, vehicles (also referred to as "connected devices" and "smart devices"), buildings and other items—embedded with electronics, software, sensors, actuators, and network connectivity that enable these objects to collect and exchange data. In 2013 the Global Standards Initiative on Internet of Things (IoT-GSI) defined the IoT as "the infrastructure of the information society." The IoT allows objects to be sensed and/or controlled remotely across existing network infrastructure, creating opportunities for more direct integration of the physical world into computer-based systems, and resulting in improved efficiency, accuracy and economic benefit. When IoT is augmented with sensors and actuators, the technology becomes an instance of the more general class of cyber-physical systems, which also encompasses technologies such as smart grids, smart homes, intelligent transportation and smart cities. Each thing is uniquely identifiable through its embedded computing system but is able to interoperate within the existing Internet infrastructure. Experts estimate that the IoT will consist of almost 50 billion objects by 2020.

Drawing representing the Internet of Things

Typically, IoT is expected to offer advanced connectivity of devices, systems, and services that goes beyond machine-to-machine (M2M) communications and covers a variety of protocols, domains, and applications. The interconnection of these embedded devices (including smart objects), is expected to usher in automation in nearly all fields, while also enabling advanced applications like a smart grid, and expanding to the areas such as smart cities.

"Things," in the IoT sense, can refer to a wide variety of devices such as heart monitoring implants, biochip transponders on farm animals, electric clams in coastal waters, automobiles with built-in sensors, DNA analysis devices for environmental/food/pathogen monitoring or field operation devices that assist firefighters in search and rescue operations. Legal scholars suggest to look at "Things" as an "inextricable mixture of hardware, software, data and service". These devices collect useful data with the help of various existing technologies and then autonomously flow the data between other devices. Current market examples include home automation (also known as smart home devices) such as the control and automation of lighting, heating (like smart thermostat), ventilation, air conditioning (HVAC) systems, and appliances such as washer/dryers, robotic vacuums, air purifiers, ovens or refrigerators/freezers that use Wi-Fi for remote monitoring.

As well as the expansion of Internet-connected automation into a plethora of new application areas, IoT is also expected to generate large amounts of data from diverse locations, with the consequent necessity for quick aggregation of the data, and an increase in the need to index, store, and process such data more effectively. IoT is one of the platforms of today's Smart City, and Smart Energy Management Systems.

The concept of the Internet of Things was invented by and term coined by Peter T. Lewis in September 1985 in a speech he delivered at a U.S. Federal Communications Commission (FCC) supported session at the Congressional Black Caucus 15th Legislative Weekend Conference.

History

As of 2016, the vision of the Internet of things has evolved due to a convergence of multiple technologies, including ubiquitous wireless communication, real-time analytics, machine learning, commodity sensors, and embedded systems. This means that the traditional fields of embedded systems, wireless sensor networks, control systems, automation (including home and building automation), and others all contribute to enabling the Internet of things (IoT).

The concept of a network of smart devices was discussed as early as 1982, with a modified Coke machine at Carnegie Mellon University becoming the first Internet-connected appliance, able to report its inventory and whether newly loaded drinks were cold. Mark Weiser's seminal 1991 paper on ubiquitous computing, "The Computer of the 21st Century", as well as academic venues such as UbiComp and PerCom produced the contemporary vision of IoT. In 1994 Reza Raji described the concept in *IEEE Spectrum* as "[moving] small packets of data to a large set of nodes, so as to integrate and automate everything from home appliances to entire factories". Between 1993 and 1996 several companies proposed solutions like Microsoft's at Work or Novell's NEST. However, only in 1999 did the field start gathering momentum. Bill Joy envisioned Device to Device (D2D) communication as part of his "Six Webs" framework, presented at the World Economic Forum at Davos in 1999.

The term "Internet of Things" was coined by Peter T. Lewis in a 1985 speech given at a U.S. Federal Communications Commission (FCC) supported wireless session at the Congressional Black Caucus 15th Legislative Weekend Conference. In his speech he states that "The Internet of Things, or IoT, is the integration of people, processes and technology with connectable devices and sensors to enable remote monitoring, status, manipulation and evaluation of trends of such devices."

The concept of the Internet of things became popular in 1999, through the Auto-ID Center at MIT and related market-analysis publications. Radio-frequency identification (RFID) was seen by Kevin Ashton (one of the founders of the original Auto-ID Center) as a prerequisite for the Internet of things at that point. Ashton prefers the phrase "Internet *for* Things." If all objects and people in daily life were equipped with identifiers, computers could manage and inventory them. Besides using RFID, the *tagging* of things may be achieved through such technologies as near field communication, barcodes, QR codes and digital watermarking.

In its original interpretation, one of the first consequences of implementing the Internet of things by equipping all objects in the world with minuscule identifying devices or machine-readable identifiers would be to transform daily life. For instance, instant and ceaseless inventory control would become ubiquitous. A person's ability to interact with objects could be altered remotely based on immediate or present needs, in accordance with existing end-user agreements. For example, such technology could grant motion-picture publishers much more control over end-user private devices by remotely enforcing copyright restrictions and digital rights management, so the ability of a customer who bought a Blu-ray disc to watch the movie becomes dependent on so-called "copyright holder's" decision, similar to Circuit City's failed DIVX.

Applications

A Nest learning thermostat reporting on energy usage and local weather

According to Gartner, Inc. (a technology research and advisory corporation), there will be nearly 20.8 billion devices on the Internet of things by 2020. ABI Research estimates that more than 30 billion devices will be wirelessly connected to the Internet of things by 2020. As per a 2014 survey and study done by Pew Research Internet Project, a large majority of the technology experts and engaged Internet users who responded—83 percent—agreed with the notion that the Internet/Cloud of Things, embedded and wearable computing (and the corresponding dynamic systems) will have widespread and beneficial effects by 2025. As such, it is clear that the IoT will consist of a very large number of devices being connected to the Internet. In an active move to accommodate new and emerging technological innovation, the UK Government, in their 2015 budget, allocated £40,000,000 towards research into the Internet of things. The former British Chancellor of the Exchequer George Osborne, posited that the Internet of things is the next stage of the information revolution and referenced the inter-connectivity of everything from urban transport to medical devices to household appliances.

A 2012 Internet refrigerator from LG

Integration with the Internet implies that devices will use an IP address as a unique identifier. However, due to the limited address space of IPv4 (which allows for 4.3 billion unique addresses), objects in the IoT will have to use IPv6 to accommodate the extremely large address space required. Objects in the IoT will not only be devices with sensory capabilities, but also provide actuation capabilities (e.g., bulbs or locks controlled over the Internet). To a large extent, the future of the Internet of things will not be possible without the support of IPv6; and consequently the global adoption of IPv6 in the coming years will be critical for the successful development of the IoT in the future.

The ability to network embedded devices with limited CPU, memory and power resources means that IoT finds applications in nearly every field. Such systems could be in charge of collecting information in settings ranging from natural ecosystems to buildings and factories, thereby finding applications in fields of environmental sensing and urban planning.

On the other hand, IoT systems could also be responsible for performing actions, not just sensing things. Intelligent shopping systems, for example, could monitor specific users' purchasing habits in a store by tracking their specific mobile phones. These users could then be provided with special offers on their favorite products, or even location of items that they need, which their fridge has automatically conveyed to the phone. Additional examples of sensing and actuating are reflected in applications that deal with heat, electricity and energy management, as well as cruise-assisting transportation systems. Other applications that the Internet of things can provide is enabling extended home security features and home automation. The concept of an "Internet of living things" has been proposed to describe networks of biological sensors that could use cloud-based analyses to allow users to study DNA or other molecules. All these advances add to the numerous list of IoT applications. Now with IoT, you can control the electrical devices installed in your house while you are sorting out your files in office. Your water will be warm as soon as you get up in the morning for the shower. All credit goes to smart devices that make up the smart home. Everything connected with the help of Internet.

However, the application of the IoT is not only restricted to these areas. Other specialized use cases of the IoT may also exist. An overview of some of the most prominent application areas is provided here. Based on the application domain, IoT products can be classified broadly into five different categories: smart wearable, smart home, smart city, smart environment, and smart enterprise. The IoT products and solutions in each of these markets have different characteristics.

Media

In order to hone the manner in which things, media and big data are interconnected, it is first necessary to provide some context into the mechanism used for media process. It has been suggested by Nick Couldry and Joseph Turow that practitioners in media approach big data as many actionable points of information about millions of individuals. The industry appears to be moving away from the traditional approach of using specific media environments such as newspapers, magazines, or television shows and instead tap into consumers with technologies that reach targeted people at optimal times in optimal locations. The ultimate aim is of course to serve, or convey, a message or content that is (statistically speaking) in line with the consumer's mindset. For example, publishing environments are increasingly tailoring the messages (advertisements) and content (articles) to appeal to consumers that have been exclusively gleaned through various data-mining activities.

The media industries process big data in a dual, interconnected manner:

- Targeting of consumers (for advertising by marketers)
- Data-capture

Thus, the Internet of things creates an opportunity to measure, collect and analyse an ever-increasing variety of behavioural statistics. Cross-correlation of this data could revolutionise the targeted marketing of products and services. For example, as noted by Danny Meadows-Klue, the combination of analytics for conversion tracking with behavioural targeting has unlocked a new level of precision that enables display advertising to be focused on the devices of people with relevant interests. Big data and the IoT work in conjunction. From a media perspective, data is the key derivative of device interconnectivity, whilst being pivotal in allowing clearer accuracy in targeting. The Internet of things therefore transforms the media industry, companies and even governments, opening up a new era of economic growth and competitiveness. The wealth of data generated by this industry (i.e. big data) will allow practitioners in advertising and media to gain an elaborate layer on the present targeting mechanisms used by the industry.

Environmental Monitoring

Environmental monitoring applications of the IoT typically use sensors to assist in environmental protection by monitoring air or water quality, atmospheric or soil conditions, and can even include areas like monitoring the movements of wildlife and their habitats. Development of resource constrained devices connected to the Internet also means that other applications like earthquake or tsunami early-warning systems can also be used by emergency services to provide more effective aid. IoT devices in this application typically span a large geographic area and can also be mobile. It has been argued that the standardization IoT brings to wireless sensing will revolutionize this area.

Infrastructure Management

Monitoring and controlling operations of urban and rural infrastructures like bridges, railway tracks, on- and offshore- wind-farms is a key application of the IoT. The IoT infrastructure can be used for monitoring any events or changes in structural conditions that can compromise safety and increase risk. It can also be used for scheduling repair and maintenance activities in an efficient manner, by coordinating tasks between different service providers and users of these facilities. IoT devices can also be used to control critical infrastructure like bridges to provide access to ships. Usage of IoT devices for monitoring and operating infrastructure is likely to improve incident management and emergency response coordination, and quality of service, up-times and reduce costs of operation in all infrastructure related areas. Even areas such as waste management can benefit from automation and optimization that could be brought in by the IoT.

Manufacturing

Network control and management of manufacturing equipment, asset and situation management, or manufacturing process control bring the IoT within the realm on industrial applications and smart manufacturing as well. The IoT intelligent systems enable rapid manufacturing of new products, dynamic response to product demands, and real-time optimization of manufacturing production and supply chain networks, by networking machinery, sensors and control systems together.

Digital control systems to automate process controls, operator tools and service information systems to optimize plant safety and security are within the purview of the IoT. But it also extends itself to asset management via predictive maintenance, statistical evaluation, and measurements to maximize reliability. Smart industrial management systems can also be integrated with the Smart Grid, thereby enabling real-time energy optimization. Measurements, automated controls, plant optimization, health and safety management, and other functions are provided by a large number of networked sensors.

National Science Foundation established an Industry/University Cooperative Research Center on intelligent maintenance systems (IMS) in 2001 with a research focus to use IoT-based predictive analytics technologies to monitor connected machines and to predict machine degradation, and further to prevent potential failures. The vision to achieve near-zero breakdown using IoT-based predictive analytics led the future development of e-manufacturing and e-maintenance activities.

The term IIOT (Industrial Internet of Things) is often encountered in the manufacturing industries, referring to the industrial subset of the IoT. IIoT in manufacturing would probably generate so much business value that it will eventually lead to the fourth industrial revolution, so the so-called Industry 4.0. It is estimated that in the future, successful companies will be able to increase their revenue through Internet of things by creating new business models and improve productivity, exploit analytics for innovation, and transform workforce. The potential of growth by implementing IIoT will generate $12 trillion of global GDP by 2030.

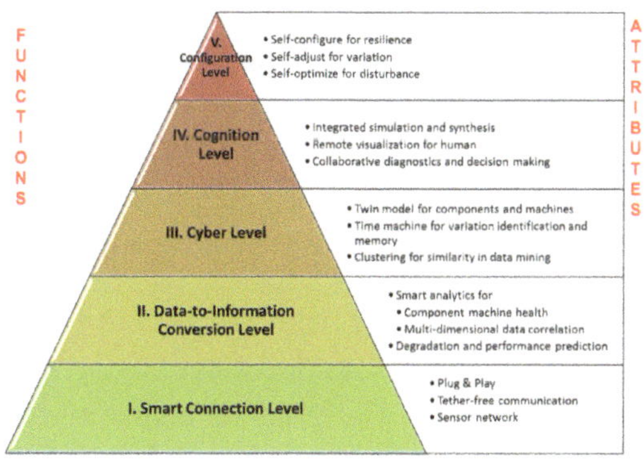

Design architecture of cyber-physical systems-enabled manufacturing system

While connectivity and data acquisition are imperative for IIoT, they should not be the purpose, rather the foundation and path to something bigger. Among all the technologies, predictive maintenance is probably a relatively "easier win" since it is applicable to existing assets and management systems. The objective of intelligent maintenance systems is to reduce unexpected downtime and increase productivity. And to realize that alone would generate around up to 30% over total maintenance costs. Industrial big data analytics will play a vital role in manufacturing asset predictive maintenance, although that is not the only capability of industrial big data. Cyber-physical systems (CPS) is the core technology of industrial big data and it will be an interface between human and the cyber world. Cyber-physical systems can be designed by following the *5C* (connection, conversion, cyber, cognition, configuration) architecture, and it will transform the collected data into actionable information, and eventually interfere with the physical assets to optimize processes.

An IoT-enabled intelligent system of such cases has been demonstrated by the NSF Industry/University Collaborative Research Center for Intelligent Maintenance Systems (IMS) at University of Cincinnati on a band saw machine in IMTS 2014 in Chicago. Band saw machines are not necessarily expensive, but the band saw belt expenses are enormous since they degrade much faster. However, without sensing and intelligent analytics, it can be only determined by experience when the band saw belt will actually break. The developed prognostics system will be able to recognize and monitor the degradation of band saw belts even if the condition is changing, so that users will know in near real time when is the best time to replace band saw. This will significantly improve user experience and operator safety, and save costs on replacing band saw belts before they actually break. The developed analytical algorithms were realized on a cloud server, and was made accessible via the Internet and on mobile devices.

Energy Management

Integration of sensing and actuation systems, connected to the Internet, is likely to optimize energy consumption as a whole. It is expected that IoT devices will be integrated into all forms of energy consuming devices (switches, power outlets, bulbs, televisions, etc.) and be able to communicate with the utility supply company in order to effectively balance power generation and energy usage. Such devices would also offer the opportunity for users to remotely control their devices, or centrally manage them via a cloud based interface, and enable advanced functions like scheduling (e.g., remotely powering on or off heating systems, controlling ovens, changing lighting conditions etc.).

Besides home based energy management, the IoT is especially relevant to the Smart Grid since it provides systems to gather and act on energy and power-related information in an automated fashion with the goal to improve the efficiency, reliability, economics, and sustainability of the production and distribution of electricity. Using advanced metering infrastructure (AMI) devices connected to the Internet backbone, electric utilities can not only collect data from end-user connections, but also manage other distribution automation devices like transformers and reclosers.

Medical and Healthcare

IoT devices can be used to enable remote health monitoring and emergency notification systems. These health monitoring devices can range from blood pressure and heart rate monitors to advanced devices capable of monitoring specialized implants, such as pacemakers Fitbit electronic wristbands or advanced hearing aids. Some hospitals have begun implementing "smart beds" that can detect when they are occupied and when a patient is attempting to get up. It can also adjust itself to ensure appropriate pressure and support is applied to the patient without the manual interaction of nurses. Specialized sensors can also be equipped within living spaces to monitor the health and general well-being of senior citizens, while also ensuring that proper treatment is being administered and assisting people regain lost mobility via therapy as well. Other consumer devices to encourage healthy living, such as, connected scales or wearable heart monitors, are also a possibility with the IoT. More and more end-to-end health monitoring IoT platforms are coming up for antenatal and chronic patients, helping one manage health vitals and recurring medication requirements.

Building and Home Automation

IoT devices can be used to monitor and control the mechanical, electrical and electronic systems used in various types of buildings (e.g., public and private, industrial, institutions, or residential) in home automation and building automation systems.

Transportation

Digital variable speed-limit sign

The IoT can assist in integration of communications, control, and information processing across various transportation systems. Application of the IoT extends to all aspects of transportation systems (i.e. the vehicle, the infrastructure, and the driver or user). Dynamic interaction between these components of a transport system enables inter and intra vehicular communication, smart traffic control, smart parking, electronic toll collection systems, logistic and fleet management, vehicle control, and safety and road assistance.

Metropolitan Scale Deployments

There are several planned or ongoing large-scale deployments of the IoT, to enable better management of cities and systems. For example, Songdo, South Korea, the first of its kind fully equipped and wired smart city, is near completion. Nearly everything in this city is planned to be wired, connected and turned into a constant stream of data that would be monitored and analyzed by an array of computers with little, or no human intervention.

Another application is a currently undergoing project in Santander, Spain. For this deployment, two approaches have been adopted. This city of 180,000 inhabitants, has already seen 18,000 city application downloads for their smartphones. This application is connected to 10,000 sensors that enable services like parking search, environmental monitoring, digital city agenda among others. City context information is used in this deployment so as to benefit merchants through a spark deals mechanism based on city behavior that aims at maximizing the impact of each notification.

Other examples of large-scale deployments underway include the Sino-Singapore Guangzhou Knowledge City; work on improving air and water quality, reducing noise pollution, and increasing transportation efficiency in San Jose, California; and smart traffic management in western Singapore. French company, Sigfox, commenced building an ultra-narrowband wireless data network in the San Francisco Bay Area in 2014, the first business to achieve such a deployment in the U.S. It subsequently announced it would set up a total of 4000 base stations to cover a total of 30 cities in the U.S. by the end of 2016, making it the largest IoT network coverage provider in the country thus far.

Another example of a large deployment is the one completed by New York Waterways in New York City to connect all the city's vessels and be able to monitor them live 24/7. The network was designed and engineered by Fluidmesh Networks, a Chicago-based company developing wireless networks for critical applications. The NYWW network is currently providing coverage on the Hudson River, East River, and Upper New York Bay. With the wireless network in place, NY Waterway is able to take control of its fleet and passengers in a way that was not previously possible. New applications can include security, energy and fleet management, digital signage, public Wi-Fi, paperless ticketing and others.

Consumer Application

A growing portion of IoT devices are created for consumer use. Examples of consumer applications include connected car, entertainment, residences and smart homes, wearable technology, quantified self, connected health, and smart retail. Consumer IoT provides new opportunities for user experience and interfaces.

Some consumer applications have been criticized for their lack of redundancy and their inconsistency, leading to a popular parody known as the "Internet of Shit." Companies have been criticized for their rush into IoT, creating devices of questionable value, and not setting up stringent security standards.

Unique Addressability of Things

The original idea of the Auto-ID Center is based on RFID-tags and unique identification through the Electronic Product Code however this has evolved into objects having an IP address or URI.

An alternative view, from the world of the Semantic Web focuses instead on making all things (not just those electronic, smart, or RFID-enabled) addressable by the existing naming protocols, such as URI. The objects themselves do not converse, but they may now be referred to by other agents, such as powerful centralized servers acting for their human owners.

The next generation of Internet applications using Internet Protocol version 6 (IPv6) would be able to communicate with devices attached to virtually all human-made objects because of the extremely large address space of the IPv6 protocol. This system would therefore be able to scale to the large numbers of objects envisaged.

A combination of these ideas can be found in the current GS1/EPCglobal EPC Information Services (EPCIS) specifications. This system is being used to identify objects in industries ranging from aerospace to fast moving consumer products and transportation logistics.

Trends and Characteristics

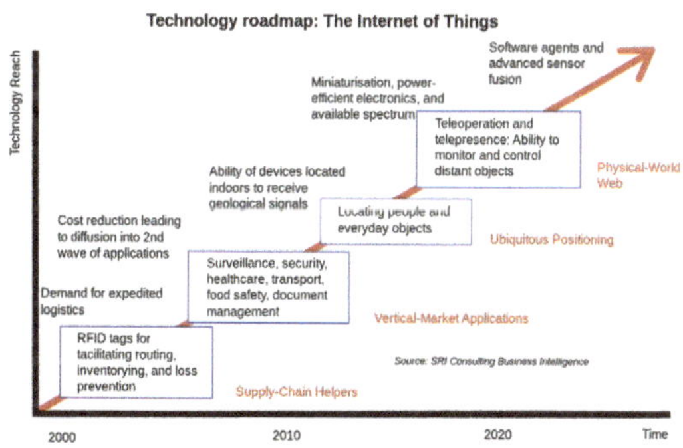

Technology roadmap: Internet of things

Intelligence

Ambient intelligence and autonomous control are not part of the original concept of the Internet of things. Ambient intelligence and autonomous control do not necessarily require Internet structures, either. However, there is a shift in research to integrate the concepts of the Internet of things and autonomous control, with initial outcomes towards this direction considering objects as the driving force for autonomous IoT.

In the future the Internet of things may be a non-deterministic and open network in which auto-organized or intelligent entities (Web services, SOA components), virtual objects (avatars) will be interoperable and able to act independently (pursuing their own objectives or shared ones) depending on the context, circumstances or environments. Autonomous behavior through the collection and reasoning of context information as well as the objects ability to detect changes in the environment, faults affecting sensors and introduce suitable mitigation measures constitute a major research trend, clearly needed to provide credibility to the IoT technology. Modern IoT products and solutions in the marketplace use a variety of different technologies to support such context-aware automation but more sophisticated forms of intelligence are requested to permit sensor units to be deployed in real environments.

Architecture

The system will likely be an example of event-driven architecture, *bottom-up* made (based on the context of processes and operations, in real-time) and will consider any subsidiary level. Therefore, model driven and functional approaches will coexist with new ones able to treat exceptions and unusual evolution of processes (multi-agent systems, B-ADSc, etc.).

In an Internet of things, the meaning of an event will not necessarily be based on a deterministic or syntactic model but would instead be based on the context of the event itself: this will also be a semantic web. Consequently, it will not necessarily need common standards that would not be able to address every context or use: some actors (services, components, avatars) will accordingly be self-referenced and, if ever needed, adaptive to existing common standards (*predicting everything* would be no more than defining a "global finality" for everything that is just not possible with any of the current *top-down* approaches and standardizations). Some researchers argue that sensor networks are the most essential components of the Internet of things.

Building on top of the Internet of things, the web of things is an architecture for the application layer of the Internet of things looking at the convergence of data from IoT devices into Web applications to create innovative use-cases. In order to program and control the flow of information in the Internet of things, a predicted architectural direction is being called BPM Everywhere which is a blending of traditional process management with process mining and special capabilities to automate the control of large numbers of coordinated devices.

Network Architecture

The Internet of things requires huge scalability in the network space to handle the surge of devices. IETF 6LoWPAN would be used to connect devices to IP networks. With billions of devices being added to the Internet space, IPv6 will play a major role in handling the network layer scalability. IETF's Constrained Application Protocol, MQTT and ZeroMQ would provide lightweight data transport.

Fog Computing is a viable alternative to prevent such large burst of data flow through Internet. The edge devices' computation power can be used to analyse and process data, thus providing easy real time scalability.

Complexity

In semi-open or closed loops (i.e. value chains, whenever a global finality can be settled) IoT will often be considered and studied as a complex system due to the huge number of different links, interactions between autonomous actors, and its capacity to integrate new actors. At the overall stage (full open loop) it will likely be seen as a chaotic environment (since systems always have finality). As a practical approach, not all elements in the Internet of things run in a global, public space. Subsystems are often implemented to mitigate the risks of privacy, control and reliability. For example, Domestic Robotics (Domotics) running inside a smart home might only share data within and be available via a local network.

Size Considerations

The Internet of things would encode 50 to 100 trillion objects, and be able to follow the movement of those objects. Human beings in surveyed urban environments are each surrounded by 1000 to 5000 trackable objects.

Space Considerations

In the Internet of things, the precise geographic location of a thing—and also the precise geographic dimensions of a thing—will be critical. Therefore, facts about a thing, such as its location in time and space, have been less critical to track because the person processing the information can decide whether or not that information was important to the action being taken, and if so, add the missing information (or decide to not take the action). (Note that some things in the Internet of things will be sensors, and sensor location is usually important.) The GeoWeb and Digital Earth are promising applications that become possible when things can become organized and connected by location. However, the challenges that remain include the constraints of variable spatial scales, the need to handle massive amounts of data, and an indexing for fast search and neighbor operations. In the Internet of things, if things are able to take actions on their own initiative, this human-centric mediation role is eliminated. Thus, the time-space context that we as humans take for granted must be given a central role in this information ecosystem. Just as standards play a key role in the Internet and the Web, geospatial standards will play a key role in the Internet of things.

Sectors

There are three core sectors of the IoT: enterprise, home, and government, with the Enterprise Internet of Things (EIoT) being the largest of the three. By 2019, the EIoT sector is estimated to account for nearly 40% or 9.1 billion devices.

A "Basket of Remotes"

According to the CEO of Cisco, the commercial opportunity for "connected products ranging from cars to household goods" is expected to be a $USD 19 trillion. Many IoT devices have a potential to take a piece of this market. Jean-Louis Gassée (Apple initial alumni team, and BeOS co-founder) has addressed this topic in an article on *Monday Note*, where he predicts that the most likely problem will be what he calls the "basket of remotes" problem, where we'll have hundreds of applications to interface with hundreds of devices that don't share protocols for speaking with one another.

There are multiple approaches to solve this problem, one of them called the "predictive interaction", where cloud or fog based decision makers will predict the user's next action and trigger some reaction.

For user interaction, new technology leaders are joining forces to create standards for communication between devices. While AllJoyn alliance is composed of the top 20 World technology leaders, there are also big companies that promote their own protocol like CCF from Intel.

Manufacturers are becoming more conscious of this problem, and many companies have begun releasing their devices with open APIs. Many of these APIs are used by smaller companies looking to take advantage of quick integration.

Frameworks

IoT frameworks might help support the interaction between "things" and allow for more complex structures like distributed computing and the development of distributed applications. Currently, some IoT frameworks seem to focus on real time data logging solutions like Jasper Technologies, Inc. and Xively (formerly Cosm and before that Pachube), offering some basis to work with many "things" and have them interact. Future developments might lead to specific software-development environments to create the software to work with the hardware used in the Internet of things. Companies are developing technology platforms to provide this type of functionality for the Internet of things. Newer platforms are being developed, which add more intelligence. Foremost, IBM has announced cognitive IoT, which combines traditional IoT with machine intelligence and learning, contextual information, industry-specific models and even natural language processing. The XMPP Standards Foundation (XSF) is creating such a framework in a fully open standard that is neither tied to any company nor connected to any cloud services. This XMPP initiative is called Chatty Things. XMPP provides a set of needed building blocks and a proven distributed solution that can scale with high security levels.

REST is a scalable architecture that allows things to communicate over Hypertext Transfer Protocol and is easily adopted for IoT applications to provide communication from a thing to a central web server. MQTT is a publish-subscribe architecture on top of TCP/IP which allows bi-directional communication between a thing and a MQTT broker.

Standards and Standards Organizations

This is a list of technical standards for the IoT, most of which are open standards, and the standards organizations that aspire to successfully setting them.

Short name	Long name	Standards under development	Other notes
Auto-ID Labs	–	Networked RFID (radiofrequency identification) and emerging sensing technologies	
EPCglobal	–	Standards for adoption of EPC (Electronic Product Code) technology	
FDA	U.S. Food and Drug Administration	UDI (Unique Device Identification) system for unique identifiers for medical devices	
GS1	–	Standards for UIDs (unique identifiers) and RFID of fast-moving consumer goods (consumer packaged goods), health care supplies, and other things	Parent organization comprises member organizations such as GS1 US
IETF	Internet Engineering Task Force	Standards that comprise TCP/IP (the Internet protocol suite)	
MTConnect Institute	–	MTConnect is a manufacturing industry standard for data exchange with machine tools and related industrial equipment. It is important to the IIoT subset of the IoT.	
OCF	Open Connectivity Foundation	Standards for simple devices using CoAP (Constrained Application Protocol)	OCF (Open Connectivity Foundation) supersedes OIC (Open Interconnect Consortium)
XSF	XMPP Standards Foundation	Protocol extensions of XMPP (Extensible Messaging and Presence Protocol), the open standard of instant messaging	

Enabling Technologies for the IOT

There are many technologies that enable IOT. A key foundation piece is the network used to communicate between nodes of an IOT installation, a role that several wireless and/or wired technologies may fulfill:

1. RFID and near-field communication – In the 2000s, RFID was the dominant technology. Later, NFC became dominant. NFC have become common in smartphones during the early 2010s, with uses such as reading NFC tags or for access to public transportation.
2. Rapid developments in the Optical technologies like Li-Fi, Cisco's 40 Gbit/s bidirectional optical technology (BiDi) may aid the development of IoT.
3. Optical tags and quick response codes – This is used for low cost tagging. Phone cameras decode QR code using image-processing techniques. In reality QR advertisement campaigns gives less turnout as users need to have another application to read QR codes.
4. Bluetooth low energy – This is one of the latest tech. All newly releasing smartphones have BLE hardware in them. Tags based on BLE can signal their presence at a power budget that enables them to operate for up to one year on a lithium coin cell battery.
5. Low energy wireless IP networks – embedded radio in system-on-a-chip designs, lower power WiFi, sub-GHz radio in an ISM band, often using a compressed version of IPv6 called 6LowPAN and relying on dedicated routing protocols such as LOADng standardized by the ITU under the recommendation ITU-T G.9903 and RPL.
6. ZigBee – This communication technology is based on the IEEE 802.15.4 2.4 GHz-band radio protocol to implement physical and MAC layer for low-rate wireless Private Area Networks. Some of its main characteristics like low power consumption, low data rate, low cost, and high message throughput make it an interesting IoT enabler technology.
7. Z-Wave – is a communication protocol that is mostly used in smart home applications. It uses a radio protocol in the 900 MHz-band.
8. Thread – Like ZigBee, this IoT communication technology relies on the IEEE 802.15.4 2.4 GHz-band radio protocol. A key difference is that its networking protocol is IPv6-compatible.
9. LTE-Advanced – LTE-A is a high-speed communication specification for mobile networks. Compared to its original LTE, LTE-A has been improved to have extended coverage, higher throughput and lower latency. One important application of this technology is Vehicle-to-Vehicle (V2V) communications.
10. WiFi-Direct – It is essentially WiFi for peer-to-peer communication without needing to have an access point. This feature attracts IoT applications to be built on top of WiFi-Direct to get benefit from the speed of WiFi while they experience lower latency.

11. HaLow - long range, low-power WiFi. Isn't able to transfer much data (bandwidth), yet sufficient for simple devices that require a network connection
12. HomePlug – This networking standard can be used to enable IOT communication over a home or building's power lines
13. MoCA – This networking standard can be used to enable IOT communication over CATV-type coaxial cable
14. Ethernet – This general purpose networking standard can be used to enable IOT communication over twisted pair or fiber network links

Simulation

IOT modeling and simulation (and emulation) is typically carried out at the design stage before deployment of the network. Network simulators like OPNET, NetSim and NS2 can be used to simulate IOT networks.

Politics and Civic Engagement

Some scholars and activists argue that the IoT can be used to create new models of civic engagement if device networks can be open to user control and inter-operable platforms. Philip N. Howard, a professor and author, writes that political life in both democracies and authoritarian regimes will be shaped by the way the IoT will be used for civic engagement. For that to happen, he argues that any connected device should be able to divulge a list of the "ultimate beneficiaries" of its sensor data, and that individual citizens should be able to add new organizations to the beneficiary list. In addition, he argues that civil society groups need to start developing their IoT strategy for making use of data and engaging with the public.

Government Regulation on IoT

One of the key drivers of the IoT is data. The success of the idea of connecting devices to make them more efficient is dependent upon access to and storage & processing of data. For this purpose, companies working on IoT collect data from multiple sources and store it in their cloud network for further processing. This leaves the door wide open for privacy and security dangers and single point vulnerability of multiple systems. The other issues pertain to consumer choice and ownership of data and how it is used. Presently the regulators have shown more interest in protecting the first three issues identified above.

Current regulatory environment:

A report published by the Federal Trade Commission (FTC) in January 2015 made the following three recommendations:

- Data security – At the time of designing IoT companies should ensure that data collection, storage and processing would be secure at all times. Companies should adopt a "defence in depth" approach and encrypt data at each stage.
- Data consent – users should have a choice as to what data they share with IoT companies and the users must be informed if their data gets exposed.
- Data minimization – IoT companies should collect only the data they need and retain the collected information only for a limited time.

However, the FTC stopped at just making recommendations for now. According to an FTC analysis, the existing framework, consisting of the FTC Act, the Fair Credit Reporting Act, and the Children's Online Privacy Protection Act, along with developing consumer education and business guidance, participation in multi-stakeholder efforts and advocacy to other agencies at the federal, state and local level, is sufficient to protect consumer rights.

A resolution passed by the Senate in March 2015, is already being considered by the Congress. This resolution recognized the need for formulating a National Policy on IoT and the matter of privacy, security and spectrum . Furthermore, to provide an impetus to the IoT ecosystem, in March 2016, a bipartisan group of four Senators proposed a bill, The Developing Innovation and Growing the Internet of Things (DIGIT) Act, to direct the Federal Communications Commission to assess the need for more spectrum to connect IoT devices.

Several standards for the IoT industry are actually being established relating to automobiles because most concerns arising from use of connected cars apply to healthcare devices as well. In fact, the National Highway Traffic Safety Administration (NHTSA) is preparing cybersecurity guidelines and a database of best practices to make automotive computer systems more secure.

Criticism and Controversies

Platform Fragmentation

IoT suffers from platform fragmentation and lack of technical standards a situation where the variety of IoT devices, in terms of both hardware variations and differences in the software running on them, makes the task of developing applications that work consistently between different inconsistent technology ecosystems hard. Customers may be hesitant to bet their IoT future on a proprietary software or hardware devices that uses proprietary protocols that may fade or become difficult to customize and interconnect.

IoT's amorphous computing nature is also a problem for security, since patches to bugs found in the core operating system often do not reach users of older and lower-price devices. One set of researchers say that the failure of vendors to support older devices with patches and updates leaves more than 87% of active devices vulnerable.

Privacy, Autonomy and Control

Philip N. Howard, a professor and author, writes that the Internet of things offers immense potential for empowering citizens, making government transparent, and broadening information access. Howard cautions, however, that privacy threats are enormous, as is the potential for social control and political manipulation.

Concerns about privacy have led many to consider the possibility that big data infrastructures such as the Internet of things and Data Mining are inherently incompatible with privacy. Writer Adam Greenfield claims that these technologies are not only an invasion of public space but are also being used to perpetuate normative behavior, citing an instance of billboards with hidden cameras that tracked the demographics of passersby who stopped to read the advertisement.

The Internet of Things Council compared the increased prevalence of digital surveillance due to the Internet of things to the conceptual panopticon described by Jeremy Bentham in the 18th Century. The assertion was defended by the works of French philosophers Michel Foucault and Gilles Deleuze. In *Discipline and Punish: The Birth of the Prison* Foucault asserts that the panopticon was a central element of the discipline society developed during the Industrial Era. Foucault also argued that the discipline systems established in factories and school reflected Bentham's vision of panopticism. In his 1992 paper "Postscripts on the Societies of Control," Delueze wrote that the discipline society had transitioned into a control society, with the computer replacing the panopticon as an instrument of discipline and control while still maintaining the qualities similar to that of panopticism.

The privacy of households could be compromised by solely analyzing smart home network traffic patterns without dissecting the contents of encrypted application data, yet a synthetic packet injection scheme can be used to safely overcome such invasion of privacy.

Peter-Paul Verbeek, a professor of philosophy of technology at the University of Twente, Netherlands, writes that technology already influences our moral decision making, which in turn affects human agency, privacy and autonomy. He cautions against viewing technology merely as a human tool and advocates instead to consider it as an active agent.

Justin Brookman, of the Center for Democracy and Technology, expressed concern regarding the impact of IoT on consumer privacy, saying that "There are some people in the commercial space who say, 'Oh, big data — well, let's collect everything, keep it around forever, we'll pay for somebody to think about security later.' The question is whether we want to have some sort of policy framework in place to limit that."

Tim O'Reilly believes that the way companies sell the IoT devices on consumers are misplaced, disputing the notion that the IoT is about gaining efficiency from putting all kinds of devices online and postulating that "IoT is really about human augmentation. The applications are profoundly different when you have sensors and data driving the decision-making."

Editorials at WIRED have also expressed concern, one stating "What you're about to lose is your privacy. Actually, it's worse than that. You aren't just going to lose your privacy, you're going to have to watch the very concept of privacy be rewritten under your nose."

The American Civil Liberties Union (ACLU) expressed concern regarding the ability of IoT to erode people's control over their own lives. The ACLU wrote that "There's simply no way to forecast how these immense powers – disproportionately accumulating in the hands of corporations seeking financial advantage and governments craving ever more control – will be used. Chances are big data and the Internet of Things will make it harder for us to control our own lives, as we grow increasingly transparent to powerful corporations and government institutions that are becoming more opaque to us."

Researchers have identified privacy challenges faced by all stakeholders in IoT domain, from the manufacturers and app developers to the consumers themselves, and examined the responsibility of each party in order to ensure user privacy at all times. Problems highlighted by the report include:

- User consent – somehow, the report says, users need to be able to give informed consent to data collection. Users, however, have limited time and technical knowledge.
- Freedom of choice – both privacy protections and underlying standards should promote freedom of choice.
- Anonymity – IoT platforms pay scant attention to user anonymity when transmitting data, the researchers note. Future platforms could, for example, use TOR or similar technologies so that users can't be too deeply profiled based on the behaviors of their "things".

In response to rising concerns about privacy and smart technology, in 2007 the British Government stated it would follow formal Privacy by Design principles when implementing their smart metering program. The program would lead to replacement of traditional power meters with smart power meters, which could track and manage energy usage more accurately. However the British Computer Society is doubtful these principles were ever actually implemented. In 2009 the Dutch Parliament rejected a similar smart metering program, basing their decision on privacy concerns. The Dutch program later revised and passed in 2011.

Security

Concerns have been raised that the Internet of things is being developed rapidly without appropriate consideration of the profound security challenges involved and the regulatory changes that might be necessary. According to the Business Insider Intelligence Survey conducted in the last quarter of 2014, 39% of the respondents said that security is the biggest concern in adopting Internet of things technology. In particular, as the Internet of things spreads widely, cyber attacks are likely to become an increasingly physical (rather than simply virtual) threat. In a January 2014 article in *Forbes*, cybersecurity columnist Joseph Steinberg listed many Internet-connected appliances that can already "spy on people in their own homes" including televisions, kitchen appliances, cameras, and thermostats. Computer-controlled devices in automobiles such as brakes, engine, locks, hood and truck releases, horn, heat, and dashboard have been shown to be vulnerable to attackers who have access to the onboard network. In some cases, vehicle computer systems are Internet-connected, allowing them to be exploited remotely. By 2008 security researchers had shown the ability to remotely control pacemakers without authority. Later hackers demonstrated remote control of insulin pumps and implantable cardioverter defibrillators. David Pogue wrote that some recently published reports about hackers remotely controlling certain functions of automobiles were not as serious as one might otherwise guess because of various mitigating circumstances; such as the bug that allowed the hack having been fixed before the report was published, or that the hack required security researchers having physical access to the car prior to the hack to prepare for it.

The U.S. National Intelligence Council in an unclassified report maintains that it would be hard to deny "access to networks of sensors and remotely-controlled objects by enemies of the United States, criminals, and mischief makers... An open market for aggregated sensor data could serve the interests of commerce and security no less than it helps criminals and spies identify vulnerable targets. Thus, massively parallel sensor fusion may undermine social cohesion, if it proves to be fundamentally incompatible with Fourth-Amendment guarantees against unreasonable search." In general, the intelligence community views the Internet of things as a rich source of data.

As a response to increasing concerns over security, the Internet of Things Security Foundation (IoTSF) was launched on 23 September 2015. IoTSF has a mission to secure the Internet of things by promoting knowledge and best practice. Its founding board is made from technology providers and telecommunications companies including BT, Vodafone, Imagination Technologies and Pen Test Partners.

In 2016, a distributed denial of service attack powered by Internet of things devices running the Mirai malware took down a DNS provider and major web sites.

Design

Given widespread recognition of the evolving nature of the design and management of the Internet of things, sustainable and secure deployment of IoT solutions must design for "anarchic scalability." Application of the concept of anarchic scalability can be extended to physical systems (i.e. controlled real-world objects), by virtue of those systems being designed to account for uncertain management futures. This "hard anarchic scalability" thus provides a pathway forward to fully realize the potential of Internet of things solutions by selectively constraining physical systems to allow for all management regimes without risking physical failure.

Brown University computer scientist Michael Littman has argued that successful execution of the Internet of things requires consideration of the interface's usability as well as the technology itself. These interfaces need to be not only more user-friendly but also better integrated: "If users need to learn different interfaces for their vacuums, their locks, their sprinklers, their lights, and their coffeemakers, it's tough to say that their lives have been made any easier."

Environmental Sustainability Impact

A concern regarding Internet of things technologies pertains to the environmental impacts of the manufacture, use, and eventual disposal of all these semiconductor-rich devices. Modern electronics are replete with a wide variety of heavy metals and rare-earth metals, as well as highly toxic synthetic chemicals. This makes them extremely difficult to properly recycle. Electronic components are often incinerated or placed in regular landfills. Furthermore, the human and environmental cost of mining the rare-earth metals that are integral to modern electronic components continues to grow. With production of electronic equipment growing globally yet little of the metals (from end-of-life equipment) are being recovered for reuse, the environmental impacts can be expected to increase.

Also, because the concept of Internet of things entails adding electronics to mundane devices (for example, simple light switches), and because the major driver for replacement of electronic components is often technological obsolescence rather than actual failure to function, it is reasonable to expect that items that previously were kept in service for many decades would see an accelerated replacement cycle if they were part of the IoT. For example, a traditional house built with 30 light switches and 30 electrical outlets might stand for 50 years, with all those components still original at the end of that period. But a modern house built with the same number of switches and outlets set up for IoT might see each switch and outlet replaced at five-year intervals, in order to keep up to date with technological changes. This translates into a ten-fold increase in waste requiring disposal.

Intentional Obsolescence of Devices

The Electronic Frontier Foundation has raised concerns that companies can use the technologies necessary to support connected devices to intentionally disable or "brick" their customers' devices via a remote software update or by disabling a service necessary to the operation of the device. In one example, home automation devices sold with the promise of a "Lifetime Subscription" were rendered useless after Nest Labs acquired Revolv and made the decision to shut down the central servers the Revolv devices had used to operate. As Nest is a company owned by Alphabet (Google's parent company), the EFF argues this sets a "terrible precedent for a company with ambitions to sell self-driving cars, medical devices, and other high-end gadgets that may be essential to a person's livelihood or physical safety."

Owners should be free to point their devices to a different server or collaborate on improved software. But such action violates the United States DMCA section 1201, which only has an exemption for "local use". This forces tinkerers who want to keep using their own equipment into a legal grey area. EFF thinks buyers should refuse electronics and software that prioritize the manufacturer's wishes above their own.

Examples of post-sale manipulations include Google Nest Revolv, disabled privacy settings on Android, Sony disabling GNU/Linux on PlayStation 3, enforced EULA on Wii U.

Business Process Automation

Business process automation (BPA), is the strategy a business uses to automate processes in order to contain costs. It consists of integrating applications, restructuring labor resources and using software applications throughout the organization.

Delivery

There are three main techniques for delivering automation of a business process.

Extension of Existing IT Systems

As most IT systems are inherently automation engines in themselves, a valid option is to extend their functionality to enable the desired automation, creating customized linkages between the disparate application systems where needed. This approach means that the automation can be tailored specifically to the exact environment of the organization; on the down-side, it can be time-consuming to find the necessary skills either internally or in the marketplace.

The boundaries of the concept end with the boundaries of the vision. If vision does not have boundaries the concept does not have boundaries (source?).

Purchase of a Specialist BPA Tool

Specialist companies are now bringing toolsets to market which are purpose-built for the function of BPA. These companies tend to focus on different industry sectors but their underlying approach tends to be similar in that they will attempt to provide the shortest route to automation by exploiting the user interface layer rather than going deeply into the application code or databases sitting behind them. They also simplify their own interface to the extent that these tools can be used directly by non-technically qualified staff. The main advantage of these toolsets is therefore their speed of deployment, the drawback is that it brings yet another IT supplier to the organization.

The market is however evolving in this area. In order to automate these processes, connectors are needed to fit these systems/solutions together with a data exchange layer to transfer the information. A Process Driven Messaging Service is an option for optimizing your data exchange layer. By mapping your end-to-end process workflow, you can build an integration between individual platforms using a Process Driven Messaging Platform. Process Driven Messaging Service gives you the logic to build your process by using triggers, jobs and workflows. Some companies uses an API where you build workflow/s and then connect various systems or mobile devices. You build the process, creating workflows in the API where the workflow in the API acts as a data exchange layer.

Purchase of a Business Process Management Solution with BPA Extensions

The next section explains that a Business Process Management system is quite different from BPA. However, it is possible to build automation on the back of a BPM implementation. The actual tools to achieve this vary, from writing custom application code to using specialist BPA tools, as described above. The advantages and disadvantages of this approach are inextricably linked – the BPM implementation provides an architecture for all processes in the business to be mapped, but this in itself delays the automation of individual processes and so benefits may be lost in the meantime.

Business Process Automation (BPA) v. Business Process Engineering (BPE) v. Business Process Management (BPM)

An area of discussion exists as to whether BPA is a distinct field of activity in its own right or merely a subset of a wider activity. Given the similarity in terminology it is not surprising most casual observers would believe them to be closely related if not identical. However, to experts in these areas they carry very distinct meanings, even if they are ultimately complementary concepts. To explain this further it is necessary to define the scope of each activity:

The BPE rationale is that before any process can be automated, it is necessary to define (often at a very strategic level or enterprise-wide) all of the business processes running inside an organisation. From this the processes can be re-defined and where necessary optimised, including automation.

BPM practice asserts that before any process can be automated, it is necessary to define all of the business processes running inside the domain of endeavour. From this the processes can be re-defined and where necessary optimised, including automation.

The BPA approach states that until a process is automated, there is no real value in analysing and defining it, and that the cycle of business change is so rapid there simply isn't time to define every process before choosing which ones to address with automation, and that delivering immediate benefits creates more value.

BPE is an approach which has ceased to be of value due to the utopian nature of the practice. Process improvement methodologies such as Lean manufacturing and Six Sigma appear to align well with the BPM and BPA view of the world, as they constantly look for incremental opportunities to make processes more efficient and reduce defects, however these methodologies can also be used downstream of a BPM deployment.

Business Process Customization

Business Process Customization is a function in process management software, such as ERP, OA and ITIL, to allow enterprises to customize their work-flows in different software in order to make the software adapt to companies' existing work-flows.

Definition

Business Process Customization is the act of defining the whole work-flow from the start to closure in software.

Method

Process Customization is achieved through JBPM (Java Business Process Management), which is an open source, flexible and extensible executable work-flow framework. It takes graphical process descriptions as input. A process is composed of tasks that are connected with sequence flows. Processes represent an execution flow. The graphical diagram (flow chart) of a process is used as the basis for the communication between non-technical users and developers.

When drawing the graphical diagram (flow chart), start and ending points should be always composed before tasks.

Related Software

jBPM is a work-flow engine widely used by software companies for Process Customization. Some business process software such as Ahoova ITIL/ITSM/BSM and SAP (ERP) contain this function.

Here is an example of customizing business process in Ahoova ITIL/ITSM/BSM software to demonstrate this concept.

Users can customize business processes with Ahoova's web process designer. It takes only three steps to complete business process customization:

1. Draw processes accordingly with tools in Ahoova's web process designer;

2. Define process' attributes;

3. Choose desired customized process for each business module, ie: request, problem, change, etc.

Business Process Preservation

The field of Business Process Preservation is concerned with digitally preserving business processes and services to ensure their long-term continued access. Built on feasibility and cost-benefit analysis it analyses and recommends which aspects of a business process should be preserved and how to preserve them. It develops methodologies and tools to capture and formalise business processes on both technical and organisational levels. This includes their underlying software and hardware infrastructures and dependencies on third-party services and information. It aligns digital preservation with well-established methods for enterprise risk management and business continuity management.

Motivation

The commercial imperative for business process preservation comes from several pressures. Heavily regulated industries, such as pharmaceuticals and aircraft manufacture must fully document processes so that they can be audited, reproduced, or diagnosed. Long-lived companies must manage services across multiple changes in technical environments. Organisations that use escrow services to mitigate risk must be confident that all of the needed information is demonstrably included in the escrow agreement and services. Organisations undergoing major staff changes must ensure that they retain the knowledge needed to operate or re-instate production processes. In addition to publications and data, academics need the software and process information to assess the validity of the data and the derived scientific claims. The same provenance information that can provide a key in regulated industries can also support credit assignment in academia. All industries benefit from analysis of processes that may lead to their continuous improvement. Memory institutions need to document the provenance of their digital collections as they undergo format-shifts to prove the authenticity or quality of their process products.

Approach

The approach used by the EU co-funded TIMBUS project breaks business process preservation down into three functions:

1. Planning: performs risk analysis and determines the requirements for preserving the relevant business processes.
2. Preservation: preserves the business processes.
3. Redeployment: reactivates and reruns the business processes.

Professional Open Source

Professional open source is an open-source software business model where an open-source software vendor generates revenue from paid professional services, maintenance and support provided along with the software. Some open-source software vendors also provide commercial licenses of open-source software or customized versions of open-source software to specific customers . Professional open source is more commonly used in the business environment rather than for individual use. *"The more mission-critical the open source software, the more necessary it is to acquire paid support"*, suggested SUSE's Gerald Pfeifer. *"Individual users will often tough out solving problems through community help forums, but SMB owners and enterprise users more likely will opt for paid support rather than devoting internal resources to support open source software,"* he added.

History

As open-source software began to become popular in the 1990s with the introduction of Linux, business hesitated to adopt it because of fears that no single entity guaranteed its stability or support. As a consequence, larger businesses would often choose commercially distributed software over a product that was released under an open-source license. However, there has been much growth in the number of professional open-source companies – made popular by companies like Liferay, Inc., eXo Platform, OpenSearchServer, Red Hat, MySQL AB, and JBoss. The business model of these companies tries "to offer open-source software with a free license, while using professional services, maintenance and support for these products to derive revenue."

Professional Services for Business Use

Under the category of professional open source, there are many different types of support available to users. Some may need assistance with their personal use of particular open source software, in which case there are options such as forums, live chat rooms and video tutorials available for free. In the case of business use however, there are more paid professional services available.

Technical Support

Technical support is one of the professional services available to users of open source software, this type of support can include a wide range of services from bug repairs to simple usage guidance. The case with most major open source software packages for business use is that the vendor of the software gives clients the option of commercial support in addition to the free software. Businesses can pay for this type of service on a contractual basis with regular monthly payments. Alternatively, they can receive technical support through independent firms.

Software Training Classes and Workshops

In order for business operations to run efficiently, it is important for members of staff, who will be using the open source software, to be very familiar with the functionalities of the software. Businesses in this situation are likely to benefit from organising for their staff to be trained before they use the software. This can be carried out by sending staff to training classes and workshops provided by independent firms such as Open Logic. To prevent any interruptions of work, similar training is available online through a virtual learning environment provided by firms such as CBT Nuggets.

Assistance with Customisation

A lot of companies using open source software for business are using it due to the convenient ability they have to adapt it to their own business needs. *"A starting point for modifying open source applications is taking an open source solution that comes close to your needs and adding your own modifications from there"*, suggested Mac McConnell, vice president of Bonitasoft. *"Companies that still do their own software have very specific needs. Most companies today want to start with open source foundation and then customize from there."* This may, however, not be possible without the assistance of someone with high technical knowledge of customizing the source code. To overcome this problem, businesses can hire professional assistance from the software developers such as QAT Global.

Approaches

A dominant factor contributing to the preference of proprietary software as opposed to open-source software is the common misconception that there is more support available for the former than the latter. Developers of open-source software are under no licensing agreement to provide any technical support for their software, nor is any particular organisation responsible for doing so. However, professional open source provides technical support for users through one of the following two approaches and is a prime example of how users can counter this issue.

Part of the Package

Certain vendors of open source software such as Red Hat include technical support as part of the package they provide. Staying with Red Hat as an example, they offer three levels of support for their version of enterprise Linux, the first of which is self-support which is simply the open source software with no support for the user. Secondly, they offer a standard level which includes support via the Internet and telephone for an unlimited number of cases during business hours. Red Hat also offer a premium level of support which would be provided at a much higher price but would provide support to the user 24 hours a day, 7 days a week for severe cases of technical difficulties. Red Hat is not the only company providing support as part of the software package, other companies with similar strategies include Canonical, AT&T, Rackspace, SUSE, HP, IBM, Nebula and many more.

Optional Extra Support

Another approach which is used by open source software vendors when providing users of their software with technical support is to provide it as an optional extra. An example of a company offering this service is Ubuntu, which provides an open source operating system for a range of devices. Customers also have the option of subscribing to Ubuntu Advantage, which is their professional support package.

Open Source Benefits

As the industry grows, the amount of open source software available is increasing. Of the present day, there are free alternative (open source) options available for almost every type of proprietary software. As well as the wide range of availability, there are several reasons why an individual user or a business model may benefit from using open source software instead of expensive proprietary options.

The number of software bugs found in the source code of open source software is gradually decreasing, and in some cases it has been found that proprietary software contains more bugs than open source software which has been built for the same purpose. A common reason for this is the fact that the source code is editable by anyone who wants access to it, therefore software developers are able to remove any bugs and errors as soon as they come across them.

The obvious advantage of using open source software in business models is the monetary savings that can be made through doing so. As there are no direct costs involved with using open source software, it can be said that businesses can save fortunes through not having to pay large amounts of money for costly proprietary software. In 2008 the Standish Group reported a drop in revenue of $60 billion in the propriety software industry due to the adoption of open source software. This also suggests that users had saved a combined $60 billion through using open source software instead of proprietary software.

A feature of open source software which has made it more favourable than proprietary software is the ability to modify the code to suit the needs of a particular user. An example of this was when Linux was used by NASA for their opsLAN system. *"We migrated key functions from Windows to Linux because we needed an operating system that was stable and reliable – one that would give us in-house control. So if we needed to patch, adjust or adapt, we could,"* said Keith Chuvala of the United Space Alliance.

Reasons for Favouring Proprietary Software

Alternatively to professional open source software, businesses may look into proprietary versions. There are several reasons why this may be a better option despite the mentioned benefits of open source software.

Open source software systems are not sold to generate profit, therefore they tend not to be particularly well suited for all end users and are often developed to suit highly technical users with similar IT backgrounds as the developer of the software. Therefore, factors such as ease of use and cleanliness of the user interface are often overlooked.

External support for open source software often comes at a cost, whereas some proprietary software packages are provided with free technical support. Similar support for open source software is available but very scarce and, when found, likely to require a monetary payment.

Despite the fact that the source code being open to the public means that users can eliminate bugs or generally improve the software, it also means that malicious users can access the code and, in theory, infect the source code with further bugs and errors.

Ultimately the businesses needs and current affairs would need to be assessed in order to determine whether proprietary software or open source software would be the most beneficial.

Alternative Support

As well as the support offered from the vendors of professional open source software packages, there are alternative methods of receiving support for open source software.

Free communal support is available on-line for most open source software packages. Mozilla Firefox, for instance, offers a service called 'Army of Awesome'. This support system allows users to post their queries to Mozilla through Twitter, the queries can then be answered by contributors. Other packages such as Linux have on-line forums dedicated to individuals and small businesses. Similarly to 'Army of Awesome', Linux users can post their queries to the dedicated forum and wait for a response from another member of the forum.

Independent IT consultant firms can also provide technical support for certain open source software. Some consultant firms, such as Unicon and OpenLogic specialise solely on open source software and can provide assistance with a range of technical difficulties. Some firms will charge a fee for this service whereas others provide the service for free.

Virtual Business Model

Virtual Business Model, is a way to organize an innovative startup company and facilitates increased flexibility in the utilization of both financial and human resources and can promote development of new ideas and inventions.

In the Virtual Company, the utilization of the financial resources can be optimized with cost-effective product development as a result. This business model is defined using several criteria; the company has a limited number of employees; the management has competence for product and business development; the company has financial resources to perform or has the ambition to find such financial resources; the company has a defined plan for the use of the financial resources; the majority of the operations are performed at organizations (called External Resources Provider) outside the Virtual Company; the ownership of the created value (e.g. technical results, patents) developed by the External Resources Providers belongs to the Virtual Company.

After the foundation of the Virtual Company and the development of the business using External Resources Providers, the company can continue to use the Virtual Company format for continued product development or after some time the company can transform the business to a traditional integrated company.

References

- [The Complete Business Process Handbook Volume 1: 'Body of Knowledge from Process Modeling to BPM by Prof. August-Wilhelm Scheer, Henrik von Scheel, Prof. Mark von Rosing, et al. (Morgan Kaufmann, ISBN 0128028602).
- Vitasek, Kate, et. al. (2011). The Vested Outsourcing Manual (1st ed.). New York: Palgrave Macmillan. ISBN 0230112684.
- Vermesan, Ovidiu; Friess, Peter (2013). Internet of Things: Converging Technologies for Smart Environments and Integrated Ecosystems (PDF). Aalborg, Denmark: River Publishers. ISBN 978-87-92982-96-4.

- Höller, J.; Tsiatsis, V.; Mulligan, C.; Karnouskos, S.; Avesand, S.; Boyle, D. (2014). From Machine-to-Machine to the Internet of Things: Introduction to a New Age of Intelligence. Elsevier. ISBN 978-0-12-407684-6.
- Fickas, S.; Kortuem, G.; Segall, Z. (13–14 Oct 1997). "Software organization for dynamic and adaptable wearable systems". International Symposium on Wearable Computers: 56–63. doi:10.1109/ISWC.1997.629920. ISBN 0-8186-8192-6.
- Waldner, Jean-Baptiste (2008). Nanocomputers and Swarm Intelligence. London: ISTE. pp. 227–231. ISBN 1-84704-002-0.
- Miles, Stephen B. (2011). RFID Technology and Applications. London: Cambridge University Press. pp. 6–8. ISBN 978-0-521-16961-5.
- Waldner, Jean-Baptiste (2007). Nanoinformatique et intelligence ambiante. Inventer l'Ordinateur du XXIeme Siècle. London: Hermes Science. p. 254. ISBN 2-7462-1516-0.
- Howard, Philip N. (2015). Pax Technica: How the internet of things May Set Us Free, Or Lock Us Up. New Haven, CT: Yale University Press. ISBN 978-0-30019-947-5.
- Verbeek, Peter-Paul (2011). Moralizing Technology: Understanding and Designing the Morality of Things. Chicago: The University of Chicago Press. ISBN 978-0-22685-291-1.
- Loukas, George (June 2015). Cyber-Physical Attacks A growing invisible threat. Oxford, UK: Butterworh-Heinemann (Elsevier). p. 65. ISBN 9780128012901.
- Karl M. Popp (2015). Best Practices for commercial use of open source software. Norderstedt, Germany: BOD. ISBN 978-3738619096.
- Karl M. Popp and Ralf Meyer (2010). Profit from Software Ecosystems: Business Models, Ecosystems and Partnerships in the Software Industry. Norderstedt, Germany: BOD. ISBN 3-8391-6983-6.
- "Internet of Things: Science Fiction or Business Fact?" (PDF). Harvard Business Review. November 2014. Retrieved 23 October 2016.
- "An Introduction to the Internet of Things (IoT)" (PDF). Cisco.com. San Francisco, California: Lopez Research. November 2013. Retrieved 23 October 2016.
- Mattern, Friedemann; Floerkemeier, Christian. "From the Internet of Computers to the Internet of Things" (PDF). ETH Zurich. Retrieved 23 October 2016.
- Reddy, Aala Santhosh (May 2014). "Reaping the Benefits of the Internet of Things" (PDF). Cognizant. Retrieved 23 October 2016.

5

Business Intelligence: An Integrated Study

Business intelligence is a theory that includes the strategies and applications and also the technologies that are used in any business. It helps in easy interpretations of large volumes of data. The aspects elucidated in this section are of vital importance, and provides a better understanding of business intelligence.

Business Intelligence

Business intelligence (BI) is a philosophy which includes the strategies, processes, applications, data, products, technologies and technical architectures used to support the collection, analysis, presentation and dissemination of business information. BI technologies are capable of handling large amounts of structured and sometimes unstructured data to help identify, develop and otherwise create new strategic business opportunities. The goal of BI is to allow for the easy interpretation of these large volumes of data. Identifying new opportunities and implementing an effective strategy based on insights can provide businesses with a competitive market advantage and long-term stability.

BI technologies provide historical, current and predictive views of business operations. Common functions of business intelligence technologies are reporting, online analytical processing, analytics, data mining, process mining, complex event processing, business performance management, benchmarking, text mining, predictive analytics and prescriptive analytics.

BI can be used to support a wide range of business decisions ranging from operational to strategic. Basic operating decisions include product positioning or pricing. Strategic business decisions include priorities, goals and directions at the broadest level. In all cases, BI is most effective when it combines data derived from the market in which a company operates (external data) with data from company sources internal to the business such as financial and operations data (internal data). When combined, external and internal data can provide a more complete picture which, in effect, creates an "intelligence" that cannot be derived by any singular set of data. Amongst myriad uses, BI tools empower organisations to gain insight into new markets, assess demand and suitability of products and services for different market segments and gauge the impact of marketing efforts.

Components

Business intelligence is made up of an increasing number of components including:

- Multidimensional aggregation and allocation
- Denormalization, tagging and standardization
- Realtime reporting with analytical alert
- A method of interfacing with unstructured data sources
- Group consolidation, budgeting and rolling forecasts
- Statistical inference and probabilistic simulation
- Key performance indicators optimization
- Version control and process management
- Open item management

History

The earliest known use of the term "Business Intelligence" is in Richard Millar Devens' in the 'Cyclopædia of Commercial and Business Anecdotes' from 1865. Devens used the term to describe how the banker, Sir Henry Furnese, gained profit by receiving and acting upon information about his environment, prior to his competitors. "*Throughout Holland, Flanders, France, and Germany, he maintained a complete and perfect train of business intelligence. The news of the many battles fought was thus received first by him, and the fall of Namur added to his profits, owing to his early receipt of the news.*" (Devens, (1865), p. 210). The ability to collect and react accordingly based on the information retrieved, an ability that Furnese excelled in, is today still at the very heart of BI.

In a 1958 article, IBM researcher Hans Peter Luhn used the term business intelligence. He employed the Webster's dictionary definition of intelligence: "the ability to apprehend the interrelationships of presented facts in such a way as to guide action towards a desired goal."

Business intelligence as it is understood today is said to have evolved from the decision support systems (DSS) that began in the 1960s and developed throughout the mid-1980s. DSS originated in the computer-aided models created to assist with decision making and planning. From DSS, data warehouses, Executive Information Systems, OLAP and business intelligence came into focus beginning in the late 80s.

In 1989, Howard Dresner (later a Gartner analyst) proposed "business intelligence" as an umbrella term to describe "concepts and methods to improve business decision making by using fact-based support systems." It was not until the late 1990s that this usage was widespread.

Data Warehousing

Often BI applications use data gathered from a data warehouse (DW) or from a data mart, and the concepts of BI and DW sometimes combine as "BI/DW" or as "BIDW". A data warehouse contains a copy of analytical data that facilitates decision support. However, not all data warehouses serve for business intelligence, nor do all business intelligence applications require a data warehouse.

To distinguish between the concepts of business intelligence and data warehouses, Forrester Research defines business intelligence in one of two ways:

1. Using a broad definition: "Business Intelligence is a set of methodologies, processes, architectures, and technologies that transform raw data into meaningful and useful information used to enable more effective strategic, tactical, and operational insights and decision-making." Under this definition, business intelligence also includes technologies such as data integration, data quality, data warehousing, master-data management, text- and content-analytics, and many others that the market sometimes lumps into the "Information Management" segment. Therefore, Forrester refers to *data preparation* and *data usage* as two separate but closely linked segments of the business-intelligence architectural stack.

2. Forrester defines the narrower business-intelligence market as, "...referring to just the top layers of the BI architectural stack such as reporting, analytics and dashboards."

Comparison with Competitive Intelligence

Though the term business intelligence is sometimes a synonym for competitive intelligence (because they both support decision making), BI uses technologies, processes, and applications to analyze mostly internal, structured data and business processes while competitive intelligence gathers, analyzes and disseminates information with a topical focus on company competitors. If understood broadly, business intelligence can include the subset of competitive intelligence.

Comparison with Business Analytics

Business intelligence and business analytics are sometimes used interchangeably, but there are alternate definitions. One definition contrasts the two, stating that the term business intelligence refers to collecting business data to find information primarily through asking questions, reporting, and online analytical processes. Business analytics, on the other hand, uses statistical and quantitative tools for explanatory and predictive modeling.

In an alternate definition, Thomas Davenport, professor of information technology and management at Babson College argues that business intelligence should be divided into querying, reporting, Online analytical processing (OLAP), an "alerts" tool, and business analytics. In this definition, business analytics is the subset of BI focusing on statistics, prediction, and optimization, rather than the reporting functionality.

Applications in an Enterprise

Business intelligence can be applied to the following business purposes, in order to drive business value.

1. Measurement – program that creates a hierarchy of performance metrics and benchmarking that informs business leaders about progress towards business goals (business process management).
2. Analytics – program that builds quantitative processes for a business to arrive at optimal decisions and to perform business knowledge discovery. Frequently involves: data mining, process mining, statistical analysis, predictive analytics, predictive modeling, business process modeling, data lineage, complex event processing and prescriptive analytics.
3. Reporting/enterprise reporting – program that builds infrastructure for strategic reporting to serve the strategic management of a business, not operational reporting. Frequently involves data visualization, executive information system and OLAP.
4. Collaboration/collaboration platform – program that gets different areas (both inside and outside the business) to work together through data sharing and electronic data interchange.
5. Knowledge management – program to make the company data-driven through strategies and practices to identify, create, represent, distribute, and enable adoption of insights and experiences that are true business knowledge. Knowledge management leads to learning management and regulatory compliance.

In addition to the above, business intelligence can provide a pro-active approach, such as alert functionality that immediately notifies the end-user if certain conditions are met. For example, if some business metric exceeds a pre-defined threshold, the metric will be highlighted in standard reports, and the business analyst may be alerted via e-mail or another monitoring service. This end-to-end process requires data governance, which should be handled by the expert.

Prioritization of Projects

It can be difficult to provide a positive business case for business intelligence initiatives, and often the projects must be prioritized through strategic initiatives. BI projects can attain higher prioritization within the organization if managers consider the following:

- As described by Kimball the BI manager must determine the tangible benefits such as eliminated cost of producing legacy reports.
- Data access for the entire organization must be enforced. In this way even a small benefit, such as a few minutes saved, makes a difference when multiplied by the number of employees in the entire organization.
- As described by Ross, Weil & Roberson for Enterprise Architecture, managers should also consider letting the BI project be driven by other business initiatives with excellent business cases. To support this approach, the organization must have enterprise architects who can identify suitable business projects.
- Using a structured and quantitative methodology to create defensible prioritization in line with the actual needs of the organization, such as a weighted decision matrix.

Success Factors of Implementation

According to Kimball et al., there are three critical areas that organizations should assess before getting ready to do a BI project:

1. The level of commitment and sponsorship of the project from senior management.
2. The level of business need for creating a BI implementation.
3. The amount and quality of business data available.

Business Sponsorship

The commitment and sponsorship of senior management is according to Kimball *et al.*, the most important criteria for assessment. This is because having strong management backing helps overcome shortcomings elsewhere in the project. However, as Kimball *et al.* state: "even the most elegantly designed DW/BI system cannot overcome a lack of business [management] sponsorship".

It is important that personnel who participate in the project have a vision and an idea of the benefits and drawbacks of implementing a BI system. The best business sponsor should have organizational clout and should be well connected within the organization. It is ideal that the business sponsor is demanding but also able to be realistic and supportive if the implementation runs into delays or drawbacks. The management sponsor also needs to be able to assume accountability and to take responsibility for failures and setbacks on the project. Support from multiple members of the management ensures the project does not fail if one person leaves the steering group. However, having many managers work together on the project can also mean that there are several different interests that attempt to pull the project in different directions, such as if different departments want to put more emphasis on their usage. This issue can be countered by an early and specific analysis of the business areas that benefit the most from the implementation. All stakeholders in the project should participate in this analysis in order for them to feel invested in the project and to find common ground.

Another management problem that may be encountered before the start of an implementation is an overly aggressive business sponsor. Problems of scope creep occur when the sponsor requests data sets that were not specified in the original planning phase.

Business Needs

Because of the close relationship with senior management, another critical thing that must be assessed before the project begins is whether or not there is a business need and whether there is a clear business benefit by doing the implementation. The needs and benefits of the implementation are sometimes driven by competition and the need to gain an advantage in the market. Another reason for a business-driven approach to implementation of BI is the acquisition of other organizations that enlarge the original organization it can sometimes be beneficial to implement DW or BI in order to create more oversight.

Companies that implement BI are often large, multinational organizations with diverse subsidiaries. A well-designed BI solution provides a consolidated view of key business data not available anywhere else in the organization, giving management visibility and control over measures that otherwise would not exist.

Amount and Quality of Available Data

Without proper data, or with too little quality data, any BI implementation fails; it does not matter how good the management sponsorship or business-driven motivation is. Before implementation it is a good idea to do data profiling. This analysis identifies the "content, consistency and structure [..]" of the data. This should be done as early as possible in the process and if the analysis shows that data is lacking, put the project on hold temporarily while the IT department figures out how to properly collect data.

When planning for business data and business intelligence requirements, it is always advisable to consider specific scenarios that apply to a particular organization, and then select the business intelligence features best suited for the scenario.

Often, scenarios revolve around distinct business processes, each built on one or more data sources. These sources are used by features that present that data as information to knowledge workers, who subsequently act on that information. The business needs of the organization for each business process adopted correspond to the essential steps of business intelligence. These essential steps of business intelligence include but are not limited to:

1. Go through business data sources in order to collect needed data
2. Convert business data to information and present appropriately
3. Query and analyze data
4. Act on the collected data

The quality aspect in business intelligence should cover all the process from the source data to the final reporting. At each step, the quality gates are different:

1. Source Data:
   - Data Standardization: make data comparable (same unit, same pattern...)
   - Master Data Management: unique referential
2. Operational Data Store (ODS):
   - Data Cleansing: detect & correct inaccurate data
   - Data Profiling: check inappropriate value, null/empty
3. Data warehouse:
   - Completeness: check that all expected data are loaded
   - Referential integrity: unique and existing referential over all sources
   - Consistency between sources: check consolidated data vs sources
4. Reporting:
   - Uniqueness of indicators: only one share dictionary of indicators
   - Formula accuracy: local reporting formula should be avoided or checked

User Aspect

Some considerations must be made in order to successfully integrate the usage of business intelligence systems in a company. Ultimately the BI system must be accepted and utilized by the users in order for it to add value to the organization. If the usability of the system is poor, the users may become frustrated and spend a considerable amount of time figuring out how to use the system or may not be able to really use the system. If the system does not add value to the users´ mission, they simply don't use it.

To increase user acceptance of a BI system, it can be advisable to consult business users at an early stage of the DW/BI lifecycle, for example at the requirements gathering phase. This can provide an insight into the business process and what the users need from the BI system. There are several methods for gathering this information, such as questionnaires and interview sessions.

When gathering the requirements from the business users, the local IT department should also be consulted in order to determine to which degree it is possible to fulfill the business's needs based on the available data.

Taking a user-centered approach throughout the design and development stage may further increase the chance of rapid user adoption of the BI system.

Besides focusing on the user experience offered by the BI applications, it may also possibly motivate the users to utilize the system by adding an element of competition. Kimball suggests implementing a function on the Business Intelligence portal website where reports on system usage can be found. By doing so, managers can see how well their departments are doing and compare themselves to others and this may spur them to encourage their staff to utilize the BI system even more.

In a 2007 article, H. J. Watson gives an example of how the competitive element can act as an incentive. Watson describes how a large call centre implemented performance dashboards for all call agents, with monthly incentive bonuses tied to performance metrics. Also, agents could compare their performance to other team members. The implementation of this type of performance measurement and competition significantly improved agent performance.

BI chances of success can be improved by involving senior management to help make BI a part of the organizational culture, and by providing the users with necessary tools, training, and support. Training encourages more people to use the BI application.

Providing user support is necessary to maintain the BI system and resolve user problems. User support can be incorporated in many ways, for example by creating a website. The website should contain great content and tools for finding the necessary information. Furthermore, helpdesk support can be used. The help desk can be manned by power users or the DW/BI project team.

BI Portals

A Business Intelligence portal (BI portal) is the primary access interface for Data Warehouse (DW) and Business Intelligence (BI) applications. The BI portal is the user's first impression of the DW/BI system. It is typically a browser application, from which the user has access to all the individual services of the DW/BI system, reports and other analytical functionality. The BI portal must be implemented in such a way that it is easy for the users of the DW/BI application to call on the functionality of the application.

The BI portal's main functionality is to provide a navigation system of the DW/BI application. This means that the portal has to be implemented in a way that the user has access to all the functions of the DW/BI application.

The most common way to design the portal is to custom fit it to the business processes of the organization for which the DW/BI application is designed, in that way the portal can best fit the needs and requirements of its users.

The BI portal needs to be easy to use and understand, and if possible have a look and feel similar to other applications or web content of the organization the DW/BI application is designed for (consistency).

The following is a list of desirable features for web portals in general and BI portals in particular:

Usable

> User should easily find what they need in the BI tool.

Content Rich

> The portal is not just a report printing tool, it should contain more functionality such as advice, help, support information and documentation.

Clean

> The portal should be designed so it is easily understandable and not over-complex as to confuse the users

Current

> The portal should be updated regularly.

Interactive

> The portal should be implemented in a way that makes it easy for the user to use its functionality and encourage them to use the portal. Scalability and customization give the user the means to fit the portal to each user.

Value Oriented

> It is important that the user has the feeling that the DW/BI application is a valuable resource that is worth working on.

Marketplace

There are a number of business intelligence vendors, often categorized into the remaining independent "pure-play" vendors and consolidated "megavendors" that have entered the market through a recent trend of acquisitions in the BI industry. The business intelligence market is gradually growing. In 2012 business intelligence services brought in $13.1 billion in revenue.

Some companies adopting BI software decide to pick and choose from different product offerings (best-of-breed) rather than purchase one comprehensive integrated solution (full-service).

Industry-specific

Specific considerations for business intelligence systems have to be taken in some sectors such as governmental banking regulations or healthcare. The information collected by banking institutions and analyzed with BI software must be protected from some groups or individuals, while being fully available to other groups or individuals. Therefore, BI solutions must be sensitive to those needs and be flexible enough to adapt to new regulations and changes to existing law.

Semi-structured or Unstructured Data

Businesses create a huge amount of valuable information in the form of e-mails, memos, notes from call-centers, news, user groups, chats, reports, web-pages, presentations, image-files, video-files, and marketing material and news. According to Merrill Lynch, more than 85% of all business information exists in these forms. These information types are called either *semi-structured* or *unstructured* data. However, organizations often only use these documents once.

The managements of semi-structured data is recognized as a major unsolved problem in the information technology industry. According to projections from Gartner (2003), white collar workers spend anywhere from 30 to 40 percent of their time searching, finding and assessing unstructured data. BI uses both structured and unstructured data, but the former is easy to search, and the latter contains a large quantity of the information needed for analysis and decision making. Because of the difficulty of properly searching, finding and assessing unstructured or semi-structured data, organizations may not draw upon these vast reservoirs of information, which could influence a particular decision, task or project. This can ultimately lead to poorly informed decision making.

Therefore, when designing a business intelligence/DW-solution, the specific problems associated with semi-structured and unstructured data must be accommodated for as well as those for the structured data.

Unstructured Data Vs. Semi-structured Data

Unstructured and semi-structured data have different meanings depending on their context. In the context of relational database systems, unstructured data cannot be stored in predictably ordered columns and rows. One type of unstructured data is typically stored in a BLOB (binary large object), a catch-all data type available in most relational database management systems. Unstructured data may also refer to irregularly or randomly repeated column patterns that vary from row to row within each file or document.

Many of these data types, however, like e-mails, word processing text files, PPTs, image-files, and video-files conform to a standard that offers the possibility of metadata. Metadata can include information such as author and time of creation, and this can be stored in a relational database. Therefore, it may be more accurate to talk about this as semi-structured documents or data, but no specific consensus seems to have been reached.

Unstructured data can also simply be the knowledge that business users have about future business trends. Business forecasting naturally aligns with the BI system because business users think of their business in aggregate terms. Capturing the business knowledge that may only exist in the minds of business users provides some of the most important data points for a complete BI solution.

Problems with Semi-structured or Unstructured Data

There are several challenges to developing BI with semi-structured data. According to Inmon & Nesavich, some of those are:

1. Physically accessing unstructured textual data – unstructured data is stored in a huge variety of formats.
2. Terminology – Among researchers and analysts, there is a need to develop a standardized terminology.
3. Volume of data – As stated earlier, up to 85% of all data exists as semi-structured data. Couple that with the need for word-to-word and semantic analysis.
4. Searchability of unstructured textual data – A simple search on some data, e.g. apple, results in links where there is a reference to that precise search term. (Inmon & Nesavich, 2008) gives an example: "a search is made on the term felony. In a simple search, the term felony is used, and everywhere there is a reference to felony, a hit to an unstructured document is made. But a simple search is crude. It does not find references to crime, arson, murder, embezzlement, vehicular homicide, and such, even though these crimes are types of felonies."

The Use of Metadata

To solve problems with searchability and assessment of data, it is necessary to know something about the content. This can be done by adding context through the use of metadata. Many systems already capture some metadata (e.g. filename, author, size, etc.), but more useful would be metadata about the actual content – e.g. summaries, topics, people or companies mentioned. Two technologies designed for generating metadata about content are automatic categorization and information extraction.

Future

A 2009 paper predicted these developments in the business intelligence market:

- Because of lack of information, processes, and tools, through 2012, more than 35 percent of the top 5,000 global companies regularly fail to make insightful decisions about significant changes in their business and markets.
- By 2012, business units will control at least 40 percent of the total budget for business intelligence.
- By 2012, one-third of analytic applications applied to business processes will be delivered through coarse-grained application mashups.
- BI has a huge scope in Entrepreneurship however majority of new entrepreneurs ignore its potential.

A 2009 *Information Management* special report predicted the top BI trends: "green computing, social networking services, data visualization, mobile BI, predictive analytics, composite applications, cloud computing and multitouch". Research undertaken in 2014 indicated that employees are more likely to have access to, and more likely to engage with, cloud-based BI tools than traditional tools.

Other business intelligence trends include the following:

- Third party SOA-BI products increasingly address ETL issues of volume and throughput.
- Companies embrace in-memory processing, 64-bit processing, and pre-packaged analytic BI applications.
- Operational applications have callable BI components, with improvements in response time, scaling, and concurrency.
- Near or real time BI analytics is a baseline expectation.
- Open source BI software replaces vendor offerings.

Other lines of research include the combined study of business intelligence and uncertain data. In this context, the data used is not assumed to be precise, accurate and complete. Instead, data is considered uncertain and therefore this uncertainty is propagated to the results produced by BI.

According to a study by the Aberdeen Group, there has been increasing interest in Software-as-a-Service (SaaS) business intelligence over the past years, with twice as many organizations using this deployment approach as one year ago – 15% in 2009 compared to 7% in 2008.

An article by InfoWorld's Chris Kanaracus points out similar growth data from research firm IDC, which predicts the SaaS BI market will grow 22 percent each year through 2013 thanks to increased product sophistication, strained IT budgets, and other factors.

An analysis of top 100 Business Intelligence and Analytics scores and ranks the firms based on several open variables

Business Intelligence 2.0

Business Intelligence 2.0 (BI 2.0) is a development of the existing business intelligence model that began in the mid-2000s. The process allows for the querying of real-time corporate data by employees, but approaches the data with a more web/browser based solution. This is opposed to previous proprietary querying tools that had characterized previous business intelligence software.

The growth in service-oriented architectures (SOA) is one of the main factors for this happening. Some state that it is a more flexible and adaptive solution than normal business intelligence. The process of data exchange also differs, with XBRL (Extensible Business Reporting Language), Web Services and various Semantic Web ontologies enable using data external to an organization, such as benchmarking type information.

Business Intelligence 2.0 is believed to have been named after Web 2.0, although it takes elements from both Web 2.0 (a focus on user empowerment and community collaboration, technologies like RSS and the concept of mashups), and the Semantic Web, sometimes called "Web 3.0" (semantic integration through shared ontologies to enable easier exchange of data).

In a nutshell BI 2.0 is about the future where data can be obtained from many sources.

According to analytics expert Neil Raden, BI 2.0 also implies a move away from the standard data warehouse that business intelligence tools have used, which "will give way to context, contingency and the need to relate information quickly from many sources."

Business Analytics

Business analytics (BA) refers to the skills, technologies, practices for continuous iterative exploration and investigation of past business performance to gain insight and drive business planning. Business analytics focuses on developing new insights and understanding of business performance based on data and statistical methods. In contrast, business intelligence traditionally focuses on using a consistent set of metrics to both measure past performance and guide business planning, which is also based on data and statistical methods.

Business analytics makes extensive use of statistical analysis, including explanatory and predictive modeling, and fact-based management to drive decision making. It is therefore closely related to management science. Analytics may be used as input for human decisions or may drive fully automated decisions. Business intelligence is querying, reporting, online analytical processing (OLAP), and "alerts."

In other words, querying, reporting, OLAP, and alert tools can answer questions such as what happened, how many, how often, where the problem is, and what actions are needed. Business analytics can answer questions like why is this happening, what if these trends continue, what will happen next (that is, predict), what is the best that can happen (that is, optimize).

Examples of Application

Banks, such as Capital One, use data analysis (or *analytics*, as it is also called in the business setting), to differentiate among customers based on credit risk, usage and other characteristics and then to match customer characteristics with appropriate product offerings. Harrah's, the gaming firm, uses analytics in its customer loyalty programs. E & J Gallo Winery quantitatively analyzes and predicts the appeal of its wines. Between 2002 and 2005, Deere & Company saved more than $1 billion by employing a new analytical tool to better optimize inventory. A telecoms company that pursues efficient call centre usage over customer service may save money.

Types of Analytics

- Decision analytics: supports human decisions with visual analytics the user models to reflect reasoning.
- Descriptive analytics: gains insight from historical data with reporting, scorecards, clustering etc.
- Predictive analytics: employs predictive modeling using statistical and machine learning techniques
- Prescriptive analytics: recommends decisions using optimization, simulation, etc.

Basic Domains within Analytics

- Behavioral analytics

- Cohort Analysis
- Collections analytics
- Contextual data modeling - supports the human reasoning that occurs after viewing "executive dashboards" or any other visual analytics
- Cyber analytics
- Enterprise Optimization
- Financial services analytics
- Fraud analytics
- Marketing analytics
- Pricing analytics
- Retail sales analytics
- Risk & Credit analytics
- Supply Chain analytics
- Talent analytics
- Telecommunications
- Transportation analytics

History

Analytics have been used in business since the management exercises were put into place by Frederick Winslow Taylor in the late 19th century. Henry Ford measured the time of each component in his newly established assembly line. But analytics began to command more attention in the late 1960s when computers were used in decision support systems. Since then, analytics have changed and formed with the development of enterprise resource planning (ERP) systems, data warehouses, and a large number of other software tools and processes.

In later years the business analytics have exploded with the introduction to computers. This change has brought analytics to a whole new level and has made the possibilities endless. As far as analytics has come in history, and what the current field of analytics is today many people would never think that analytics started in the early 1900s with Mr. Ford himself.

Challenges

Business analytics depends on sufficient volumes of high quality data. The difficulty in ensuring data quality is integrating and reconciling data across different systems, and then deciding what subsets of data to make available.

Previously, analytics was considered a type of after-the-fact method of forecasting consumer behavior by examining the number of units sold in the last quarter or the last year. This type of data warehousing required a lot more storage space than it did speed. Now business analytics is becoming a tool that can influence the outcome of customer interactions. When a specific customer type is considering a purchase, an analytics-enabled enterprise can modify the sales pitch to appeal to that consumer. This means the storage space for all that data must react extremely fast to provide the necessary data in real-time.

Competing on Analytics

Thomas Davenport, professor of information technology and management at Babson College argues that businesses can optimize a distinct business capability via analytics and thus better compete. He identifies these characteristics of an organization that are apt to compete on analytics:

- One or more senior executives who strongly advocate fact-based decision making and, specifically, analytics
- Widespread use of not only descriptive statistics, but also predictive modeling and complex optimization techniques
- Substantial use of analytics across multiple business functions or processes
- Movement toward an enterprise level approach to managing analytical tools, data, and organizational skills and capabilities

Analytics

Analytics is the discovery, interpretation, and communication of meaningful patterns in data. Especially valuable in areas rich with recorded information, analytics relies on the simultaneous application of statistics, computer programming and operations research to quantify performance. Analytics often favors data visualization to communicate insight.

Organizations may apply analytics to business data to describe, predict, and improve business performance. Specifically, areas within analytics include predictive analytics, prescriptive analytics, enterprise decision management, retail analytics, store assortment and stock-keeping unit optimization, marketing optimization and marketing mix modeling, web analytics, sales force sizing and optimization, price and promotion modeling, predictive science, credit risk analysis, and fraud analytics. Since analytics can require extensive computation, the algorithms and software used for analytics harness the most current methods in computer science, statistics, and mathematics.

Analytics Vs. Analysis

Analytics is multidisciplinary. There is extensive use of mathematics and statistics, the use of descriptive techniques and predictive models to gain valuable knowledge from data—data analysis. The insights from data are used to recommend action or to guide decision making rooted in business context. Thus, analytics is not so much concerned with individual analyses or analysis steps, but with the entire methodology. There is a pronounced tendency to use the term *analytics* in business settings e.g. text analytics vs. the more generic text mining to emphasize this broader perspective.. There is an increasing use of the term *advanced analytics*, typically used to describe the technical aspects of analytics, especially in the emerging fields such as the use of machine learning techniques like neural networks to do predictive modeling.

Application of Analytics

Marketing Optimization

Marketing has evolved from a creative process into a highly data-driven process. Marketing organizations use analytics to determine the outcomes of campaigns or efforts and to guide decisions for investment and consumer targeting. Demographic studies, customer segmentation, conjoint analysis and other techniques allow marketers to use large amounts of consumer purchase, survey and panel data to understand and communicate marketing strategy.

Web analytics allows marketers to collect session-level information about interactions on a website using an operation called sessionization. Google Analytics is an example of a popular free analytics tool that marketers use for this purpose. Those interactions provide web analytics information systems with the information necessary to track the referrer, search keywords, identify IP address, and track activities of the visitor. With this information, a marketer can improve marketing campaigns, website creative content, and information architecture.

Analysis techniques frequently used in marketing include marketing mix modeling, pricing and promotion analyses, sales force optimization and customer analytics e.g.: segmentation. Web analytics and optimization of web sites and online campaigns now frequently work hand in hand with the more traditional marketing analysis techniques. A focus on digital media has slightly changed the vocabulary so that *marketing mix modeling* is commonly referred to as *attribution modeling* in the digital or marketing mix modeling context.

These tools and techniques support both strategic marketing decisions (such as how much overall to spend on marketing, how to allocate budgets across a portfolio of brands and the marketing mix) and more tactical campaign support, in terms of targeting the best potential customer with the optimal message in the most cost effective medium at the ideal time.

Portfolio Analytics

A common application of business analytics is portfolio analysis. In this, a bank or lending agency has a collection of accounts of varying value and risk. The accounts may differ by the social status (wealthy, middle-class, poor, etc.) of the holder, the geographical location, its net value, and many other factors. The lender must balance the return on the loan with the risk of default for each loan. The question is then how to evaluate the portfolio as a whole.

The least risk loan may be to the very wealthy, but there are a very limited number of wealthy people. On the other hand, there are many poor that can be lent to, but at greater risk. Some balance must be struck that maximizes return and minimizes risk. The analytics solution may combine time series analysis with many other issues in order to make decisions on when to lend money to these different borrower segments, or decisions on the interest rate charged to members of a portfolio segment to cover any losses among members in that segment.

Risk Analytics

Predictive models in the banking industry are developed to bring certainty across the risk scores for individual customers. Credit scores are built to predict individual's delinquency behavior and widely used to evaluate the credit worthiness of each applicant. Furthermore, risk analyses are carried out in the scientific world and the insurance industry. It is also extensively used in financial institutions like Online Payment Gateway companies to analyse if a transaction was genuine or fraud. For this purpose they use the transaction history of the customer. This is more commonly used in Credit Card purchase, when there is a sudden spike in the customer transaction volume the customer gets a call of confirmation if the transaction was initiated by him/her. This helps in reducing loss due to such circumstances.

Digital Analytics

Digital analytics is a set of business and technical activities that define, create, collect, verify or transform digital data into reporting, research, analyses, recommendations, optimizations, predictions, and automations. This also includes the SEO (Search Engine Optimization) where the keyword search is tracked and that data is used for marketing purposes. Even banner ads and clicks come under digital analytics. All marketing firms rely on digital analytics for their digital marketing assignments, where MROI (Marketing Return on Investment) is important.

Security Analytics

Security analytics refers to information technology (IT) solutions that gather and analyze security events to bring situational awareness and enable IT staff to understand and analyze events that pose the greatest risk. Solutions in this area include security information and event management solutions and user behavior analytics solutions.

Software Analytics

Software analytics is the process of collecting information about the way a piece of software is used and produced.

Challenges

In the industry of commercial analytics software, an emphasis has emerged on solving the challenges of analyzing massive, complex data sets, often when such data is in a constant state of change. Such data sets are commonly referred to as big data. Whereas once the problems posed by big data were only found in the scientific community, today big data is a problem for many businesses that operate transactional systems online and, as a result, amass large volumes of data quickly.

The analysis of unstructured data types is another challenge getting attention in the industry. Unstructured data differs from structured data in that its format varies widely and cannot be stored in traditional relational databases without significant effort at data transformation. Sources of unstructured data, such as email, the contents of word processor documents, PDFs, geospatial data, etc., are rapidly becoming a relevant source of business intelligence for businesses, governments and universities. For example, in Britain the discovery that one company was illegally selling fraudulent doctor's notes in order to assist people in defrauding employers and insurance companies, is an opportunity for insurance firms to increase the vigilance of their unstructured data analysis. The McKinsey Global Institute estimates that big data analysis could save the American health care system $300 billion per year and the European public sector €250 billion.

These challenges are the current inspiration for much of the innovation in modern analytics information systems, giving birth to relatively new machine analysis concepts such as complex event processing, full text search and analysis, and even new ideas in presentation. One such innovation is the introduction of grid-like architecture in machine analysis, allowing increases in the speed of massively parallel processing by distributing the workload to many computers all with equal access to the complete data set.

Analytics is increasingly used in education, particularly at the district and government office levels. However, the complexity of student performance measures presents challenges when educators try to understand and use analytics to discern patterns in student performance, predict graduation likelihood, improve chances of student success, etc. For example, in a study involving districts known for strong data use, 48% of teachers had difficulty posing questions prompted by data, 36% did not comprehend given data, and 52% incorrectly interpreted data. To combat this, some analytics tools for educators adhere to an over-the-counter data format (embedding labels, supplemental documentation, and a help system, and making key package/display and content decisions) to improve educators' understanding and use of the analytics being displayed.

One more emerging challenge is dynamic regulatory needs. For example, in the banking industry, Basel III and future capital adequacy needs are likely to make even smaller banks adopt internal risk models. In such cases, cloud computing and open source R (programming language) can help smaller banks to adopt risk analytics and support branch level monitoring by applying predictive analytics.

Risks

The main risk for the people is discrimination like price discrimination or statistical discrimination. Analytical processes can also result in discriminatory outcomes that may violate anti-discrimination and civil rights laws. There is also the risk that a developer could profit from the ideas or work done by users, like this example: Users could write new ideas in a note taking app, which could then be sent as a custom event, and the developers could profit from those ideas. This can happen because the ownership of content is usually unclear in the law.

If a user's identity is not protected, there are more risks; for example, the risk that private information about users is made public on the internet.

In the extreme, there is the risk that governments could gather too much private information, now that the governments are giving themselves more powers to access citizens' information.

Metrics Reference Model

The Metrics Reference Model (MRM) is the reference model created by the Consortium for Advanced Management-International (CAM-I) to be a single reference library of performance metrics. This library is useful for accelerating to development of and improving the content of any organization's business intelligence solution.

Background

The MRM was created by the Business Intelligence Working Group of CAM-I. Substantial secondary research performed by the group revealed that there was no single, central location for performance metrics information. All published material focused on a single dimension of performance metrics (e.g., financial), or focused on how to use such metrics to affect the performance of an organization. Exhaustive lists of performance metrics that were generic to nearly any organization across all aspects of the business (e.g., customer, product/service, employee) were not found. Furthermore, without a single reference point, no research was located where the relationships between these performance metrics had been documented and described.

Overview

MRM is a technology agnostic reference model based on the basic elements of a generic organization for which cost, process, and performance improvements may be desired.

In keeping with CAM-I's generic model of an organization, these elements include:

- Customers (or business sustaining forces)
- Products and services
- Processes and activities
- Employees
- Finance
- Research and development
- Infrastructure
- Resources
- Suppliers

The figure below illustrates the flow between these components in a typical organization.

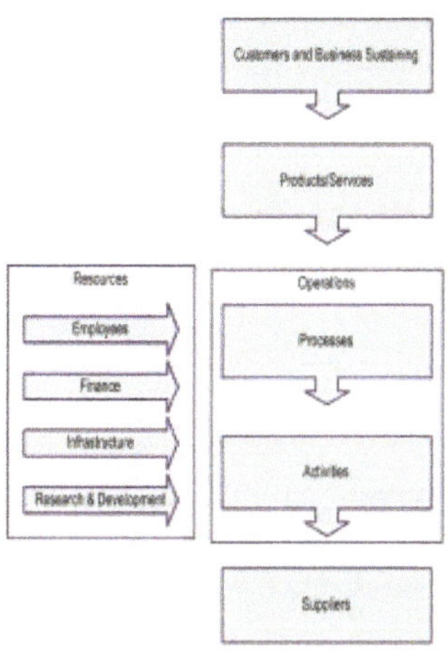

CAM-I's Generic Model of an Organization

Content

For each of the organizational elements, the MRM provides a comprehensive list of cost, process, and performance metrics of possible interest to many organizations. For example, some of the performance metrics provided for the Employee organizational element include Average Customer Review, Customer Retention Rate, Sales per Customer/Spending Rate, Average Quantity per Customer Order, Order Frequency, Average Number of Orders per Customer, and Average Customer Lifetime Value.

For each performance metric, this information is also provided:

- Type. Provides further levels of suborganization within the organizational element where sensible (e.g., Financial).
- Definition/Calculation. A generally accepted means of calculating the metric from measures is provided.
- Strengths/Opportunities. Known advantages for using the metric.
- Weaknesses/Problems/Risks. Known disadvantages for using the metric.
- Target Setting. A discussion of the desired outcome.
- Possible Correlations/Market Basket of Metrics. Consideration of other metrics and correlations of environmental factors are provided. This area explores the relationship of the measure against every other metric in the MRM, an indication of the strength of the relationship (e.g., weak, moderate, strong), and an indication to the type of relationship (e.g., negatively correlated, positively correlated).

Value and Use of the MRM

The benefit of the MRM is that each metric has already been identified, defined, and considered by a group of experienced business intelligence professionals. The idea of the MRM is for the business intelligence (BI) practitioner to follow these steps:

- Assess the requirements of the BI solution and determine which organization components apply.
- For each organizational component that applies, open this document to the associated MRM section.
- Review each of the measures.
- Identify measures of interest.
- Review the information provided about each measure and either accept or reject the measure for inclusion in the BI tool.

- Accumulate all desired measures in a list.
- Use the measures (or, if metrics are selected, use the measures in the Definition/Calculation column) to drive the BI data model.

Using the MRM in industry How the MRM may be used practically in the industry depends on whether an individual or an organization is considered.

Individual interest in the MRM:

- Program analysts: Use the MRM to tailor to a specific program/organization.
- Managers: Use the MRM model to validate that the metrics selected represent the needs of their business.
- Strategic planners: Use the MRM to determine metric relationships and correlations in order to determine strategic decisions.
- Operational planners: Use the MRM to understand operational data points that you have control over to affect the metrics and understand the components of the metrics.
- BI practitioners: Use the MRM to define the BI requirements and calculations, as well as the interpretation of the data. Help the business owner determine the appropriate display of metrics.
- Vendors: Use the MRM to provide relevant products/services that answer clients' needs out-of-the-box.
- Educators: Use the MRM to discuss the benefits and drawbacks of the various measures.

Organizational interest in the MRM:

- New organizations/agencies, mergers & acquisitions
- Existing organizations that are starting performance measurement efforts
- Consulting organizations for use in client situations
- Software vendors
- Monitoring organizations
- Those with performance challenges (may have metrics, but not enough to interpret that metric)
- Those who do not wish to go through a large strategic planning exercise but want to start using best practice metrics for common functions

- Small budgets or time constraints or those that do not have the expertise to create metrics

Use Cases The MRM to be used in a number of different ways, including:

- Use as part of planning process. Rather than starting from a blank slate, go through each and add those that are specific to their organization. * Use to generate ideas. Customize existing, add specific to your organization
- Use as a cookbook. Select ingredients (metrics) of interest from each area when there are known objectives but want to save time.
- Use to refine existing program (rare), compare metrics vs. existing (completeness review)
- BI practitioner: Use for developing the calculations, finding the data needed to run the calculations, and building appropriate data structures
- For consulting organizations to provide more thorough results quickly (competitive advantage)
- For software vendors to provide specific applications
- Education tool (understanding strengths/weaknesses of metrics, etc.) for staff at all levels to understand what's important

Financial Intelligence (Business)

Financial intelligence is the knowledge and skills gained from understanding finance and accounting principles in the business world. Although a fairly new term, financial intelligence has its roots in organizational development research, mostly in the field of employee participation. Financial intelligence has emerged as a best practice and core competency in many organizations leading to improved financial results, increased employee morale, and reduced employee turnover. Many organizations include financial intelligence programs in their leadership development curriculum. Financial intelligence is not an innate skill, rather it is a learned set of skills that can be developed at all levels.

Areas of Understanding

The four areas of understanding that make up financial intelligence are:

Understanding the foundation. Financial intelligence requires an understanding of the basics of financial measurement including the income statement, the balance sheet, and the cash flow statement. It also requires knowing the difference between cash and profit and why a balance sheet balances.

Understanding the art. Finance and accounting are an art as well as a science. The two disciplines must try to quantify what can't always be quantified, and so must rely on rules, estimates, and assumptions. Financial intelligence ensures people are able to identify where the artful aspects of finance have been applied to the numbers, and know how applying them differently might lead to different conclusions.

Understanding analysis. Financial intelligence includes the ability to analyze the numbers in greater depth. This includes being able to calculate profitability, leverage, liquidity and efficiency ratios and understanding the meaning of the results. Conducting ROI analysis and interpreting the results are also part of financial intelligence.

Understanding the big picture. Financial intelligence also means being able to understand a business's financial results in context - that is, within the framework of the big picture. Factors such as the economy, the competitive environment, regulations and changing customer needs and expectations as well as new technologies all affect how the numbers are interpreted.

Financial intelligence is not just theoretical book learning. It also requires practice and real world application. In the corporate world, managers can display financial intelligence by speaking the language, that is, asking questions about the numbers when something doesn't make sense, reviewing financial reports and using the information to understand the company's strengths and weaknesses, using ROI analysis, working capital management, and ratio analysis to make decisions, and identifying where the art of finance has been applied.

Roots of Financial Intelligence

In 1954, Peter F. Drucker, in his groundbreaking book, *The Practice of Management*, wrote the following.

"[The worker] should know how his work relates to the work of the whole. He should know what he contributes to the enterprise...if he lacks information, he will lack both incentive and means to improve his performance."

"It is in the best interest of the organization that the worker have the information."

The concept of financial intelligence in organizations comes from the research of several well-known organizational development academics, including Dennis Denison, Edward Lawler and Jeffrey Pfeffer. For example, the research of Lawler, Mohrman and Ledford found that the indices that have the most impact on both direct performance outcomes in organizations (productivity, customer satisfaction, quality and speed) and on profitability and competitiveness were sharing information and developing knowledge.

Karen Berman's research, asked specifically if information (operationalized by teaching business basics to improve financial intelligence and sharing information on a regular basis) improves the results of employee participation, as seen through organizational performance improvement and employee attitude improvement. The results of the study found that certain financial performance measures improved and that employee turnover decreased. Financial literacy also has its roots in open book management. A core tenant of open-book management is business literacy, that is, ensuring everyone understands how the business measures financial success.

Financial Intelligence in Organizations

Many companies, (Southwest Airlines being a prime example) consider financial intelligence a core competency or best practice. Several universities, including Harvard Business School, Wharton and Stanford have programs targeted at the corporate world, mostly at the leadership level, to increase the financial intelligence in organizations. There are a variety of methods to increase financial intelligence in organizations, including classroom training, webinar training, and business simulations. Training companies have supported these efforts as well, including Business Literacy Institute, Paradigm Learning, and Workplace Development.

Proponents of financial intelligence in organizations believe that if employees, managers and leaders understand financial information and how financial success is measured, they will make decisions and take action based on an understanding of the financial impact of those decisions. If everyone knows the financial goals of the company, for example, and knows how to make decisions that support those financial goals, then the company is going to be more financially successful.

Employee owned companies are one group of organizations that are focused on ensuring everyone in the company is financially intelligent, as employees are owners and therefore must understand the financial side of the business.

IBM Cognos Business Intelligence

IBM Cognos Business Intelligence is a web-based, integrated business intelligence suite by IBM. It provides a toolset for reporting, analysis, scorecarding, and monitoring of events and metrics. The software consists of several components to meet the different information requirements in a company.

Basic Components

Cognos Connection

Cognos Connection is the Web portal for IBM Cognos BI. It is the starting point for the browser-based access to all functions provided with the suite. With the help of the portal, content can be searched in the form of reports, scorecards and agents, it can be managed, structured and displayed. In addition, the portal is used for example to schedule and distribute reports, for creating tasks, administering the server and the access permissions to content available to different users. You can also create shortcuts, URLs and pages.

Query Studio

Query Studio allows simple queries and self-service reports to answer rather simple business questions. The report layout can be customized and data can be filtered and sorted, formatting and the creation of diagrams is also supported.

Report Studio

The Report Studio is used to create management reports. It offers two different modes: The professional authoring mode enables users to access the full range of Report Studio functionality. In this mode, users can create any type of report, including charts, maps, lists, and repeat functions. In professional authoring mode all types of Data (relational or multidimensional) can be used, dynamic data can not be displayed.

The express authoring mode has a more simplified user interface, designed for non-technical users. It enables them to create traditional financial or management reports in a more focused user interface. In contrast to the professional authoring mode, the express authoring mode allows the use of dynamic data.

Analysis Studio

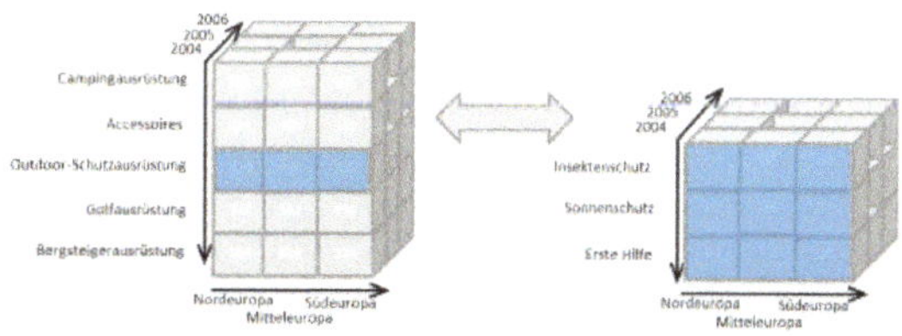

Drill-up and drill-down as example OLAP-functionalities

Users can create analyses of large data sources and search for background information about an event or action. Multidimensional analysis allows identifying trends and understanding of anomalies or deviations, which are not obvious in other types of reports. Drag-and-drop features, elements and key performance indicators can be included in the analysis, rows and columns can be switched, OLAP-functionalities like drill-up and drill-down can be used to get a deeper understanding about the sources of the information used in the analysis.

Event Studio

The Event Studio is a notification tool that informs about events within the enterprise in real time. Therefore, agents can be created to detect the occurrence of business events or exceptional circumstances, based on the change of specified event- or data conditions. A notification may be served by sending an e-mail, its publication in the portal, or by triggering reports. This can be used to handle failure with notification.

Workspace

IBM Cognos Workspace (formerly introduced in version 10.1 as IBM Cognos Business Insight and renamed in version 10.2.0) is a web-based interface that allows business users to use existing IBM Cognos content (report objects) to build interactive workspaces for insight and collaboration.

Workspace Advanced

IBM Cognos Workspace Advanced (formerly introduced in version 10.1 as IBM Cognos Business Insight Advanced and renamed in version 10.2.0) is a web-based interface that allows business users to author/create reports and analyze information.

Additional Components

Go! Office

The Go! Office component lets users work with IBM Cognos content in their familiar Microsoft Office environment. The component provides access to all IBM Cognos Report contents, including data, metadata, headers and footers, and diagrams. Users can use predefined reports or create new content with Query Studio, Analysis Studio or Report Studio. By importing content into Microsoft Excel users can use the formatting, calculation and presentation features. The created documents can then be imported using Cognos Connection, published and made available for other users.

Go! Search

In Cognos Connection, you can do a full-text search for content contained in reports, analyses, dashboards, metric information and events. When searching, an index of the prompts, titles, headings, column names, row names, data elements and other important fields is used as base. The full text search in IBM Cognos Go! Search is related to the search in regular search engines such as Google. Users can search operators such as +, - or use "" (quotation marks) to change the default behavior of search queries with multiple words. Search terms are not case sensitive, word and spelling variants are included in the results. You can also search for a specific type of entry, such as an agent. The search results are sorted in descending order, the entry with the greatest amount of relevant metadata is displayed at the top of the list. In Analysis Studio, Query Studio and IBM Cognos Viewer, you can either perform a full text search, as also search for content related to the data of the current view.

Go! Dashboard

With IBM Cognos Go! Dashboard, interactive dashboards containing IBM Cognos content and external data sources can be created to fit the information needs of an individual user.

The following items can be added to a dashboard: Report objects, they are displayed in a Cognos Viewer portlet. Report parts such as lists, crosstabs, and charts are displayed in interactive portlets. Lists or crosstabs can be displayed as a chart and vice versa. Content can be shown or hidden dynamically by the use of sliders and checkboxes. The Cognos Search portlet allows searching for published content. In addition, Web links, Web pages, RSS feeds, and images can be displayed on the dashboard.

The user interface has two modes: In the interactive mode, existing dashboards are viewed and interacted with, creating and editing of dashboards can be done in assembly mode.

Report

MOTORCYCLE ACCIDENT CAUSE FACTORS AND
IDENTIFICATION OF COUNTERMEASURES
VOLUME I: TECHNICAL REPORT

H.H. Hurt, Jr.
J.V. Ouellet
D.R. Thom

Traffic Safety Center
University of Southern California
Los Angeles, California 90007

Contract No. DOT HS-5-01160
Contract Amount $501,814.00

JANUARY 1981
FINAL REPORT

This document is available to the US. public through the National Technical Information Service, Springfield, Virginia 22161

Prepared For
U.S. DEPARTMENT OF TRANSPORTATION
National Highway Traffic Safety Administration
Washington, D.C. 20590

Example of a front page of a report

A report or account is any informational work (usually of writing, speech, television, or film) made with the specific intention of relaying information or recounting certain events in a widely presentable form.

Description

A report is an informational work made with the specific intention of relaying information or recounting certain events in a widely presentable and scrutinized form. Reports are often conveyed in writing, speech, television, or film.

Use

Reports fill a vast array of informational needs for many of society's important organizations. Reports are used for keeping track of information, which may be used to make decisions. Written reports are documents which present focused, salient content, generally to a specific audience. Reports are used in government, business, education, science, and other fields, are often to display the result of an experiment, investigation, or inquiry.

Attributes

Reports use features such as graphics, images, voice, or specialized vocabulary in order to persuade that specific audience to undertake an action. One of the most common formats for presenting reports is IMRAD: Introduction, Methods, Results and Discussion. This structure is standard for the genre because it mirrors the traditional publication of scientific research and summons the ethos and credibility of that discipline. Reports are not required to follow this pattern, and may use alternative patterns like the problem-solution format.

Additional elements often used to persuade readers include: headings to indicate topics, to more complex formats including charts, tables, figures, pictures, tables of contents, abstracts,and nouns summaries, appendices, footnotes, hyperlinks, and references.

Types

Some examples of reports are:

- annual reports
- auditor's reports
- book reports
- bound reports
- retail reports
- census reports
- credit reports
- demographic reports
- expense report
- experience report
- inspection reports

- military reports
- police reports
- policy reports
- progress reports
- recommendation reports
- investigative reports
- scientific reports
- trip reports
- white papers
- appraisal reports
- workplace reports

Collaborative Software

Collaborative software or groupware is an application software designed to help people involved in a common task to achieve their goals. One of the earliest definitions of collaborative software is "intentional group processes plus software to support them".

Collaborative software is a broad concept that overlaps considerably with computer-supported cooperative work (CSCW). According to Carstensen and Schmidt (1999) groupware is part of CSCW. The authors claim that CSCW, and thereby groupware, addresses "how collaborative activities and their coordination can be supported by means of computer systems." Software products such as email, calendaring, text chat, wiki, and bookmarking belong to this category whenever used for group work, whereas the more general term social software applies to systems used outside the workplace, for example, online dating services and social networking sites like Twitter and Facebook.

The use of collaborative software in the work space creates a collaborative working environment (CWE).

Finally, collaborative software relates to the notion of collaborative work systems, which are conceived as any form of human organization that emerges any time that collaboration takes place, whether it is formal or informal, intentional or unintentional. Whereas the groupware or collaborative software pertains to the technological elements of computer-supported cooperative work, collaborative work systems become a useful analytical tool to understand the behavioral and organizational variables that are associated to the broader concept of CSCW.

Origins

Douglas Engelbart first envisioned collaborative computing in 1951 and documented his vision in 1962, with working prototypes in full operational use by his research team by the mid-1960s, and held the first public demonstration of his work in 1968 in what is now referred to as "The Mother of All Demos." The following year, Engelbart's lab was hooked into the ARPANET, the first computer network, enabling them to extend services to a broader userbase.

Online collaborative gaming software began between early networked computer users. In 1975, Will Crowther created Colossal Cave Adventure on a DEC PDP-10 computer. As internet connections grew, so did the numbers of users and multi-user games. In 1978 Roy Trubshaw, a student at Essex University in the United Kingdom, created the game MUD (Multi-User Dungeon).

The US Government began using truly collaborative applications in the early 1990s. One of the first robust applications was the Navy's Common Operational Modeling, Planning and Simulation Strategy (COMPASS). The COMPASS system allowed up to 6 users created point-to-point connections with one another; the collaborative session only remained while at least one user stayed active, and would have to be recreated if all six logged out. MITRE improved on that model by hosting the collaborative session on a server that each user logged into. Called the Collaborative Virtual Workstation (CVW), this allowed the session to be set up in a virtual file cabinet and virtual rooms, and left as a persistent session that could be joined later.

In 1996, Pavel Curtis, who had built MUDs at PARC, created PlaceWare, a server that simulated a one-to-many auditorium, with side chat between "seat-mates", and the ability to invite a limited number of audience members to speak. In 1997, engineers at GTE used the PlaceWare engine in a commercial version of MITRE's CVW, calling it InfoWorkSpace (IWS). In 1998, IWS was chosen as the military standard for the standardized Air Operations Center. The IWS product was sold to General Dynamics and then later to Ezenia.

Groupware

Collaborative software was originally designated as *groupware* and this term can be traced as far back as the late 1980s, when Richman and Slovak (1987) wrote:

"Like an electronic sinew that binds teams together, the new *groupware* aims to place the computer squarely in the middle of communications among managers, technicians, and anyone else who interacts in groups, revolutionizing the way they work."

Even further back, in 1978 Peter and Trudy Johnson-Lenz coined the term groupware; their initial 1978 definition of groupware was, "intentional group processes plus software to support them." Later in their article they went on to explain groupware as "computer-mediated culture... an embodiment of social organization in hyperspace." Groupware integrates co-evolving human and tool systems, yet is simply a single system.

In the early 1990s the first commercial groupware products were delivered, and big companies such as Boeing and IBM started using electronic meeting systems for key internal projects. Lotus Notes appeared as a major example of that product category, allowing remote group collaboration when the internet was still in its infancy. Kirkpatrick and Losee (1992) wrote then:

"If GROUPWARE really makes a difference in productivity long term, the very definition of an office may change. You will be able to work efficiently as a member of a group wherever you have your computer. As computers become smaller and more powerful, that will mean anywhere."

Design and Implementation Issues

The complexity of groupware development is still an issue. One reason for this is the socio-technical dimension of groupware. Groupware designers do not only have to address technical issues (as in traditional software development) but also consider the organizational aspects and the social group processes that should be supported with the groupware application. Some examples for issues in groupware development are:

- Persistence is needed in some sessions. Chat and voice communications are routinely non-persistent and evaporate at the end of the session. Virtual room and online file cabinets can persist for years. The designer of the collaborative space needs to consider the information duration needs and implement accordingly.
- Authentication has always been a problem with groupware. When connections are made point-to-point, or when log-in registration is enforced, it's clear who is engaged in the session. However, audio and unmoderated sessions carry the risk of unannounced 'lurkers' who observe but do not announce themselves or contribute.
- Until recently, bandwidth issues at fixed location limited full use of the tools. These are exacerbated with mobile devices.
- Multiple input and output streams bring concurrency issues into the groupware applications.
- Motivational issues are important, especially in settings where no pre-defined group process was in place.

- Closely related to the motivation aspect is the question of reciprocity. Ellis and others have shown that the distribution of efforts and benefits has to be carefully balanced in order to ensure that all required group members really participate.

One approach for addressing these issues is the use of design patterns for groupware design. The patterns identify recurring groupware design issues and discuss design choices in a way that all stakeholders can participate in the groupware development process.

Groupware and Levels of Collaboration

Groupware can be divided into three categories depending on the level of collaboration:

1. Communication can be thought of as unstructured interchange of information. A phone call or an IM Chat discussion are examples of this.
2. Conferencing (or collaboration level, as it is called in the academic papers that discuss these levels) refers to interactive work toward a shared goal. Brainstorming or voting are examples of this.
3. Co-ordination refers to complex interdependent work toward a shared goal. A good metaphor for understanding this is to think about a sports team; everyone has to contribute the right play at the right time as well as adjust their play to the unfolding situation - but everyone is doing something different - in order for the team to win. That is complex interdependent work toward a shared goal: collaborative management.

Collaborative Management (Coordination) Tools

Collaborative management tools facilitate and manage group activities. Examples include:

- Electronic calendars (also called time management software) — schedule events and automatically notify and remind group members
- Project management systems — schedule, track, and chart the steps in a project as it is being completed
- Online proofing — share, review, approve, and reject web proofs, artwork, photos, or videos between designers, customers, and clients
- Workflow systems — collaborative management of tasks and documents within a knowledge-based business process
- Knowledge management systems — collect, organize, manage, and share various forms of information

- Enterprise bookmarking — collaborative bookmarking engine to tag, organize, share, and search enterprise data
- Prediction markets — let a group of people predict together the outcome of future events
- Extranet systems (sometimes also known as 'project extranets') — collect, organize, manage and share information associated with the delivery of a project (e.g.: the construction of a building)
- Intranet systems — team members and clients can create online documents and collaborate on projects securely
- Social software systems — organize social relations of groups
- Online spreadsheets — collaborate and share structured data and information
- Client portals — interact and share with your clients in a private online environment

Collaborative Software and Human Interaction

The design intent of collaborative software (groupware) is to transform the way documents and rich media are shared in order to enable more effective team collaboration.

Collaboration, with respect to information technology, seems to have several definitions. Some are defensible but others are so broad they lose any meaningful application. Understanding the differences in human interactions is necessary to ensure the appropriate technologies are employed to meet interaction needs.

There are three primary ways in which humans interact: conversations, transactions, and collaborations.

Conversational interaction is an exchange of information between two or more participants where the primary purpose of the interaction is discovery or relationship building. There is no central entity around which the interaction revolves but is a free exchange of information with no defined constraints generally focused on personal experiences. Communication technology such as telephones, instant messaging, and e-mail are generally sufficient for conversational interactions.

Transactional interaction involves the exchange of transaction entities where a major function of the transaction entity is to alter the relationship between participants.

In *collaborative interactions* the main function of the participants' relationship is to alter a collaboration entity (i.e., the converse of transactional).

Notable Collaborative Software Vendors

- 10,000ft
- Aconex
- Asana
- Basecamp
- Bluescape
- CEITON
- Central Desktop
- Citadel/UX
- CollaborateCloud
- Confluence
- EGroupware
- eXo Platform
- Freedcamp
- Google Docs, Sheets, and Slides
- Group-Office
- HighQ
- Huddle.net
- Hyperoffice
- IBM Connections
- Intrexx
- Jive
- Kolab
- LiquidPlanner
- Mavenlink
- OnlyOffice

- Open Text
- Pivotal Labs
- Podio
- Redbooth
- Samepage
- SharePoint
- Slack
- Smartsheet
- Team Foundation Server
- TeamWox
- Teamwork.com
- Targetprocess
- Tom's Planner
- Trello
- Yammer
- Workplace by Facebook
- Wrike
- Xait
- XWiki
- Zarafa
- Zoho Office Suite

Text Mining

Text mining, also referred to as *text data mining*, roughly equivalent to text analytics, is the process of deriving high-quality information from text. High-quality information is typically derived through the devising of patterns and trends through means such as statistical pattern learning. Text mining usually involves the process of structuring the input text (usually parsing, along with the addition of some derived linguistic features and the removal of others, and subsequent insertion into a database), deriving patterns within the structured data, and finally evaluation and interpretation of the output. 'High quality' in text mining usually refers to some combination of relevance, novelty, and interestingness. Typical text mining tasks include text categorization, text clustering, concept/entity extraction, production of granular taxonomies, sentiment analysis, document summarization, and entity relation modeling (*i.e.*, learning relations between named entities).

Text analysis involves information retrieval, lexical analysis to study word frequency distributions, pattern recognition, tagging/annotation, information extraction, data mining techniques including link and association analysis, visualization, and predictive analytics. The overarching goal is, essentially, to turn text into data for analysis, via application of natural language processing (NLP) and analytical methods.

A typical application is to scan a set of documents written in a natural language and either model the document set for predictive classification purposes or populate a database or search index with the information extracted.

Text analytics

The term text analytics describes a set of linguistic, statistical, and machine learning techniques that model and structure the information content of textual sources for business intelligence, exploratory data analysis, research, or investigation. The term is roughly synonymous with text mining; indeed, Ronen Feldman modified a 2000 description of "text mining" in 2004 to describe "text analytics." The latter term is now used more frequently in business settings while "text mining" is used in some of the earliest application areas, dating to the 1980s, notably life-sciences research and government intelligence.

The term text analytics also describes that application of text analytics to respond to business problems, whether independently or in conjunction with query and analysis of fielded, numerical data. It is a truism that 80 percent of business-relevant information originates in unstructured form, primarily text. These techniques and processes discover and present knowledge – facts, business rules, and relationships – that is otherwise locked in textual form, impenetrable to automated processing.

History

Labor-intensive manual text mining approaches first surfaced in the mid-1980s, but technological advances have enabled the field to advance during the past decade. Text mining is an interdisciplinary field that draws on information retrieval, data mining, machine learning, statistics, and computational linguistics. As most information (common estimates say over 80%) is currently stored as text, text mining is believed to have a high commercial potential value. Increasing interest is being paid to multilingual data mining: the ability to gain information across languages and cluster similar items from different linguistic sources according to their meaning.

The challenge of exploiting the large proportion of enterprise information that originates in "unstructured" form has been recognized for decades. It is recognized in the earliest definition of business intelligence (BI)

"...utilize data-processing machines for auto-abstracting and auto-encoding of documents and for creating interest profiles for each of the 'action points' in an organization. Both incoming and internally generated documents are automatically abstracted, characterized by a word pattern, and sent automatically to appropriate action points."

Yet as management information systems developed starting in the 1960s, and as BI emerged in the '80s and '90s as a software category and field of practice, the emphasis was on numerical data stored in relational databases. This is not surprising: text in "unstructured" documents is hard to process. The emergence of text analytics in its current form stems from a refocusing of research in the late 1990s from algorithm development to application, as described by Prof. Marti A. Hearst in the paper Untangling Text Data Mining:

For almost a decade the computational linguistics community has viewed large text collections as a resource to be tapped in order to produce better text analysis algorithms. In this paper, I have attempted to suggest a new emphasis: the use of large online text collections to discover new facts and trends about the world itself. I suggest that to make progress we do not need fully artificial intelligent text analysis; rather, a mixture of computationally-driven and user-guided analysis may open the door to exciting new results.

Hearst's 1999 statement of need fairly well describes the state of text analytics technology and practice a decade later.

Text Analysis Processes

Subtasks—components of a larger text-analytics effort—typically include:

- Information retrieval or identification of a corpus is a preparatory step: collecting or identifying a set of textual materials, on the Web or held in a file system, database, or content corpus manager, for analysis.
- Although some text analytics systems apply exclusively advanced statistical methods, many others apply more extensive natural language processing, such as part of speech tagging, syntactic parsing, and other types of linguistic analysis.

- Named entity recognition is the use of gazetteers or statistical techniques to identify named text features: people, organizations, place names, stock ticker symbols, certain abbreviations, and so on. Disambiguation—the use of contextual clues—may be required to decide where, for instance, "Ford" can refer to a former U.S. president, a vehicle manufacturer, a movie star, a river crossing, or some other entity.
- Recognition of Pattern Identified Entities: Features such as telephone numbers, e-mail addresses, quantities (with units) can be discerned via regular expression or other pattern matches.
- Coreference: identification of noun phrases and other terms that refer to the same object.
- Relationship, fact, and event Extraction: identification of associations among entities and other information in text
- Sentiment analysis involves discerning subjective (as opposed to factual) material and extracting various forms of attitudinal information: sentiment, opinion, mood, and emotion. Text analytics techniques are helpful in analyzing, sentiment at the entity, concept, or topic level and in distinguishing opinion holder and opinion object.
- Quantitative text analysis is a set of techniques stemming from the social sciences where either a human judge or a computer extracts semantic or grammatical relationships between words in order to find out the meaning or stylistic patterns of, usually, a casual personal text for the purpose of psychological profiling etc.

Applications

The technology is now broadly applied for a wide variety of government, research, and business needs. Applications can be sorted into a number of categories by analysis type or by business function. Using this approach to classifying solutions, application categories include:

- Enterprise Business Intelligence/Data Mining, Competitive Intelligence
- E-Discovery, Records Management
- National Security/Intelligence
- Scientific discovery, especially Life Sciences
- Sentiment Analysis Tools, Listening Platforms
- Natural Language/Semantic Toolkit or Service

- Publishing
- Automated ad placement
- Search/Information Access
- Social media monitoring

Security Applications

Many text mining software packages are marketed for security applications, especially monitoring and analysis of online plain text sources such as Internet news, blogs, etc. for national security purposes. It is also involved in the study of text encryption/decryption.

Biomedical Applications

A range of text mining applications in the biomedical literature has been described.

One online text mining application in the biomedical literature is PubGene that combines biomedical text mining with network visualization as an Internet service.

GoPubMed is a knowledge-based search engine for biomedical texts.

Software Applications

Text mining methods and software is also being researched and developed by major firms, including IBM and Microsoft, to further automate the mining and analysis processes, and by different firms working in the area of search and indexing in general as a way to improve their results. Within public sector much effort has been concentrated on creating software for tracking and monitoring terrorist activities.

Online Media Applications

Text mining is being used by large media companies, such as the Tribune Company, to clarify information and to provide readers with greater search experiences, which in turn increases site "stickiness" and revenue. Additionally, on the back end, editors are benefiting by being able to share, associate and package news across properties, significantly increasing opportunities to monetize content.

Marketing Applications

Text mining is starting to be used in marketing as well, more specifically in analytical customer relationship management. Coussement and Van den Poel (2008) apply it to improve predictive analytics models for customer churn (customer attrition).

Sentiment Analysis

Sentiment analysis may involve analysis of movie reviews for estimating how favorable a review is for a movie. Such an analysis may need a labeled data set or labeling of the affectivity of words. Resources for affectivity of words and concepts have been made for WordNet and ConceptNet, respectively.

Text has been used to detect emotions in the related area of affective computing. Text based approaches to affective computing have been used on multiple corpora such as students evaluations, children stories and news stories.

Academic Applications

The issue of text mining is of importance to publishers who hold large databases of information needing indexing for retrieval. This is especially true in scientific disciplines, in which highly specific information is often contained within written text. Therefore, initiatives have been taken such as Nature's proposal for an Open Text Mining Interface (OTMI) and the National Institutes of Health's common Journal Publishing Document Type Definition (DTD) that would provide semantic cues to machines to answer specific queries contained within text without removing publisher barriers to public access.

Academic institutions have also become involved in the text mining initiative:

- The National Centre for Text Mining (NaCTeM), is the first publicly funded text mining centre in the world. NaCTeM is operated by the University of Manchester in close collaboration with the Tsujii Lab, University of Tokyo. NaCTeM provides customised tools, research facilities and offers advice to the academic community. They are funded by the Joint Information Systems Committee (JISC) and two of the UK Research Councils (EPSRC & BBSRC). With an initial focus on text mining in the biological and biomedical sciences, research has since expanded into the areas of social sciences.
- In the United States, the School of Information at University of California, Berkeley is developing a program called BioText to assist biology researchers in text mining and analysis.

Digital Humanities and Computational Sociology

The automatic analysis of vast textual corpora has created the possibility for scholars to analyse millions of documents in multiple languages with very limited manual intervention. Key enabling technologies have been parsing, machine translation, topic categorization, and machine learning.

The automatic parsing of textual corpora has enabled the extraction of actors and their relational networks on a vast scale, turning textual data into network data. The resulting networks, which can contain thousands of nodes, are then analysed by using tools from network theory to identify the key actors, the key communities or parties, and general properties such as robustness or structural stability of the overall network, or centrality of certain nodes. This automates the approach introduced by quantitative narrative analysis, whereby subject-verb-object triplets are identified with pairs of actors linked by an action, or pairs formed by actor-object.

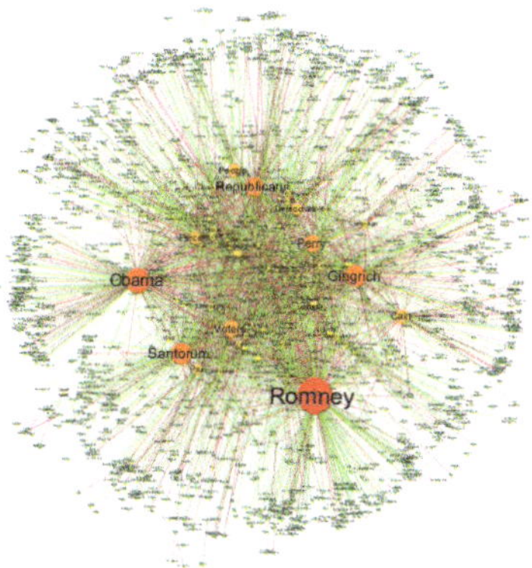

Narrative network of US Elections 2012

Content analysis has been a traditional part of social sciences and media studies for a long time. The automation of content analysis has allowed a "big data" revolution to take place in that field, with studies in social media and newspaper content that include millions of news items. Gender bias, readability, content similarity, reader preferences, and even mood have been analyzed based on text mining methods over millions of documents. The analysis of readability, gender bias and topic bias was demonstrated in Flaounas et al. showing how different topics have different gender biases and levels of readability; the possibility to detect mood shifts in a vast population by analysing Twitter content was demonstrated as well.

Software

Text mining computer programs are available from many commercial and open source companies and sources. See List of text mining software.

Intellectual Property Law

Situation in Europe

Due to a lack of flexibilities in European copyright and database law, the mining of in-copyright works such as web mining without the permission of the copyright owner is not legal. In the UK in 2014, on the recommendation of the Hargreaves review the government amended copyright law to allow text mining as a limitation and exception. Only the second country in the world to do so after Japan, which introduced a mining specific exception in 2009. However, due to the restriction of the Copyright Directive, the UK exception only allows content mining for non-commercial purposes. UK copyright law does not allow this provision to be overridden by contractual terms and conditions.

Video by Fix Copyright campaign explaining TDM and its copyright issues in the EU, 2016 [3:52]

he European Commission facilitated stakeholder discussion on text and data mining in 2013, under the title of Licences for Europe. The focus on the solution to this legal issue being licences and not limitations and exceptions to copyright law led to representatives of universities, researchers, libraries, civil society groups and open access publishers to leave the stakeholder dialogue in May 2013.

Situation in the United States

By contrast to Europe, the flexible nature of US copyright law, and in particular fair use means that text mining in America, as well as other fair use countries such as Israel, Taiwan and South Korea is viewed as being legal. As text mining is transformative, meaning that it does not supplant the original work, it is viewed as being lawful under fair use. For example, as part of the Google Book settlement the presiding judge on the case ruled that Google's digitisation project of in-copyright books was lawful, in part because of the transformative uses that the digitisation project displayed—one such use being text and data mining.

Implications

Until recently, websites most often used text-based searches, which only found documents containing specific user-defined words or phrases. Now, through use of a semantic web, text mining can find content based on meaning and context (rather than just by a specific word). Additionally, text mining software can be used to build large dossiers of information about specific people and events. For example, large datasets based on data extracted from news reports can be built to facilitate social networks analysis or counter-intelligence. In effect, the text mining software may act in a capacity similar to an intelligence analyst or research librarian, albeit with a more limited scope of analysis. Text mining is also used in some email spam filters as a way of determining the characteristics of messages that are likely to be advertisements or other unwanted material. Text mining plays an important role in determining financial market sentiment.

References

- (Rud, Olivia (2009). Business Intelligence Success Factors: Tools for Aligning Your Business in the Global Economy. Hoboken, N.J: Wiley & Sons. ISBN 978-0-470-39240-9.)
- Coker, Frank (2014). Pulse: Understanding the Vital Signs of Your Business. Ambient Light Publishing. pp. 41–42. ISBN 978-0-9893086-0-1.
- Jeanne W. Ross, Peter Weill, David C. Robertson (2006) Enterprise Architecture As Strategy, p. 117 ISBN 1-59139-839-8.
- Davenport, Thomas H.; Harris, Jeanne G. (2007). Competing on analytics : the new science of winning. Boston, Mass.: Harvard Business School Press. ISBN 978-1-4221-0332-6.
- Inmon, Bill; Nesavich, Anthony (2007). Tapping Into Unstructured Data. Prentice-Hall. ISBN 978-0-13-236029-6.
- Hearst, Marti A. (1999). "Proceedings of the 37th annual meeting of the Association for Computational Linguistics on Computational Linguistics": 3–10. doi:10.3115/1034678.1034679. ISBN 1-55860-609-2.
- Mehl, Matthias R. (2006). "Handbook of multimethod measurement in psychology": 141. doi:10.1037/11383-011. ISBN 1-59147-318-7.
- Zanasi, Alessandro (2009). "Proceedings of the International Workshop on Computational Intelligence in Security for Information Systems CISIS'08". Advances in Soft Computing. 53: 53. doi:10.1007/978-3-540-88181-0_7. ISBN 978-3-540-88180-3.
- Ananiadou, S. and McNaught, J. (Editors) (2006). Text Mining for Biology and Biomedicine. Artech House Books. ISBN 978-1-58053-984-5
- Feldman, R., and Sanger, J. (2006). The Text Mining Handbook. New York: Cambridge University Press. ISBN 978-0-521-83657-9
- Indurkhya, N., and Damerau, F. (2010). Handbook Of Natural Language Processing, 2nd Edition. Boca Raton, FL: CRC Press. ISBN 978-1-4200-8592-1
- Konchady, M. Text Mining Application Programming (Programming Series). Charles River Media. ISBN 1-58450-460-9
- Manning, C., and Schutze, H. (1999). Foundations of Statistical Natural Language Processing. Cambridge, MA: MIT Press. ISBN 978-0-262-13360-9
- Miner, G., Elder, J., Hill. T, Nisbet, R., Delen, D. and Fast, A. (2012). Practical Text Mining and Statistical Analysis for Non-structured Text Data Applications. Elsevier Academic Press. ISBN 978-0-12-386979-1

Permissions

All chapters in this book are published with permission under the Creative Commons Attribution Share Alike License or equivalent. Every chapter published in this book has been scrutinized by our experts. Their significance has been extensively debated. The topics covered herein carry significant information for a comprehensive understanding. They may even be implemented as practical applications or may be referred to as a beginning point for further studies.

We would like to thank the editorial team for lending their expertise to make the book truly unique. They have played a crucial role in the development of this book. Without their invaluable contributions this book wouldn't have been possible. They have made vital efforts to compile up to date information on the varied aspects of this subject to make this book a valuable addition to the collection of many professionals and students.

This book was conceptualized with the vision of imparting up-to-date and integrated information in this field. To ensure the same, a matchless editorial board was set up. Every individual on the board went through rigorous rounds of assessment to prove their worth. After which they invested a large part of their time researching and compiling the most relevant data for our readers.

The editorial board has been involved in producing this book since its inception. They have spent rigorous hours researching and exploring the diverse topics which have resulted in the successful publishing of this book. They have passed on their knowledge of decades through this book. To expedite this challenging task, the publisher supported the team at every step. A small team of assistant editors was also appointed to further simplify the editing procedure and attain best results for the readers.

Apart from the editorial board, the designing team has also invested a significant amount of their time in understanding the subject and creating the most relevant covers. They scrutinized every image to scout for the most suitable representation of the subject and create an appropriate cover for the book.

The publishing team has been an ardent support to the editorial, designing and production team. Their endless efforts to recruit the best for this project, has resulted in the accomplishment of this book. They are a veteran in the field of academics and their pool of knowledge is as vast as their experience in printing. Their expertise and guidance has proved useful at every step. Their uncompromising quality standards have made this book an exceptional effort. Their encouragement from time to time has been an inspiration for everyone.

The publisher and the editorial board hope that this book will prove to be a valuable piece of knowledge for students, practitioners and scholars across the globe.

Index

A

Advertising-supported Software, 75, 77

Analytics, 7, 73, 114, 168, 171, 173-174, 199, 201-202, 210-218, 235-239, 243

Artifact-centric Business Process, 89, 117-118

B

Basket of Remotes, 180

Bi Portals, 206-207

Biomedical Applications, 239

Bounty Driven Development, 74

Bpm Life-cycle, 3

Bpm Suites, 6

Bpm Technology, 7

Bpmn 2.0, 103-104

Bricks And Clicks, 49, 54-57, 128

Business Activity Monitoring, 5-6, 135, 162-163

Business Analytics, 7, 201-202, 211-214, 216

Business Intelligence, 163, 199-213, 215, 217-221, 223-225, 227, 229, 231, 233, 235-239, 241, 243

Business Intelligence 2.0, 211

Business Model, 43-61, 63-65, 67, 69, 71, 73-75, 77, 79-81, 83-85, 88-89, 91-92, 118, 130-131, 133, 135, 164, 193, 195, 197

Business Model Pattern, 58

Business Models For Open-source Software, 70, 87

Business Needs Analysis, 141, 144

Business Process, 1-44, 46, 48-50, 52-54, 56, 58, 60, 62, 64, 66, 68, 70, 72, 74, 76, 78, 80, 82-84, 86-198, 200, 202, 204, 206, 208, 210, 212, 214, 216, 218, 220, 222, 224, 226, 228, 230, 232, 234, 236, 238, 240, 242

Business Process Customization, 191-192

Business Process Integration, 91

Business Process Management, 1-42, 44, 46, 48, 50, 52, 54, 56, 58, 60, 62, 64, 66, 68, 70, 72, 74, 76, 78, 80, 82, 84, 86, 88, 90, 92-94, 96, 98, 100, 102, 104, 106, 108, 110, 112-114, 116, 118, 120, 122, 124, 126, 128, 130, 132, 134-198, 200, 202, 204, 206, 208, 210, 212, 214, 216, 218, 220, 222, 224, 226, 228, 230, 232, 234, 236, 238, 240, 242

Business Process Management (bpm), 1-2, 6, 113, 138, 163, 190

Business Process Model and Notation, 88, 90, 96, 103

Business Process Modeling, 2, 5, 53, 87-93, 95-97, 99, 101, 103, 105, 107, 109-111, 113-115, 117, 119, 121, 123, 125, 127, 129, 131, 133, 202

Business Process Preservation, 192-193

Business Process Re-engineering, 16, 113, 135

Business Process Reengineering, 6, 92-93, 111, 135-136

Business Reference Model, 52, 87, 91, 93-95

Business Sponsorship, 203

C

Cloud Computing Bpm, 8

Cognos Connection, 224, 226

Collaborative Software, 229-230, 233-234

Collective Business System, 164

Commitment-loyalty Model, 62

Competitiveness Strategy, 154

D

Data Warehousing, 201, 214

Delayed Open-sourcing, 76-77

Dependent Growth Business Model, 64

Design Content Emphasis, 48

Design Themes Emphasis, 48

Digital Analytics, 216

Digital Outsourcing, 151

Direct Selling, 50, 87, 128-129, 133

Disintermediation, 55, 125-128

Dual-licensing, 72

E

Education Strategy, 155

Effective Change Management, 141, 146

Environmental Sustainability Impact, 188

Event-driven Process Chain, 90, 119-124

F

Financial Intelligence, 222-224

Freemium, 51, 72, 75, 130-133

G

Go! Dashboard, 227

Groupware, 229-233

I
Ibm Cognos Business Intelligence, 224
Inclusive Business Model, 80-81, 83-84
Industrialization of Services Business Model, 53, 63
Intentional Obsolescence of Devices, 188
Internet of Things, 8, 135, 166-173, 177-180, 184-188, 197-198

K
Knowledge Management, 113, 202, 232

L
Loyalty Business Model, 58, 61

M
Management, 1-42, 44, 46, 48, 50, 52-54, 56-58, 60, 62-64, 66, 68, 70, 72-74, 76, 78, 80, 82, 84-90, 92-96, 98, 100, 102, 104, 106, 108-114, 116-118, 120, 122, 124, 126, 128, 130, 132, 134-206, 208-210, 212-214, 216-218, 220, 222-226, 228, 230, 232, 234, 236-240, 242
Marketing Optimization, 214-215
Mathematical Modeling, 25
Matrix Schemes, 66-67
Meta-model of Event-driven Process Chain, 124
Metrics Reference Model, 218
Multi-level Marketing, 51, 67, 70, 129

N
Network Architecture, 178

O
Obfuscation of Source Code, 76
Ongoing Continuous Improvement, 141, 147
Open Sourcing on End-of-life, 77
Operations Management, 1, 9, 13, 18-19, 25, 27-28, 35
Outsourcing, 18, 20, 44-45, 85, 111, 135, 149-162, 197

P
Platform Fragmentation, 184
Policy-making Strategy, 154
Process Chain, 90, 115, 119-124
Production Systems, 9, 16-20, 23, 25, 27, 41
Professional Open Source, 193, 195-196
Pyramid Scheme, 52, 64-65, 67-70

R
Re-licensing Under A Proprietary License, 76
Report, 29, 38, 80, 82, 116, 168, 183, 186-187, 207, 210, 225-228
Risk Analytics, 216, 218

S
Safety, Risk And Maintenance, 27
Security Analytics, 217
Security Applications, 239
Selling Professional Services, 73
Software Analytics, 217
Software Applications, 189, 239
Strategic Management, 35, 41, 53, 58, 63, 89, 109, 202
Swim Lanes, 97, 101-102, 107

T
Text Mining, 199, 215, 235-236, 239-243
The "eightball" Model, 65
The Service Quality Model, 58
Total Quality Management, 1, 15, 26, 114

V
Virtual Business Model, 135, 197
Voluntary Donations, 74

W
Workflow, 1, 3, 6, 8, 89, 95, 107, 110, 113-114, 117-120, 139, 190, 232